Patrick Moore's
Practical Astronomy Series

Springer
London
Berlin
Heidelberg
New York
Barcelona
Hong Kong
Milan
Paris
Singapore
Tokyo

Other titles in this series

The Observational Amateur Astronomer
Patrick Moore (Ed.)

Telescopes and Techniques
Chris Kitchin

The Art and Science of CCD Astronomy
David Ratledge (Ed.)

The Observer's Year
Patrick Moore

Seeing Stars
Chris Kitchin and Robert W. Forrest

Photo-guide to the Constellations
Chris Kitchin

The Sun in Eclipse
Michael Maunder and Patrick Moore

Software and Data for Practical Astronomers
David Ratledge

Amateur Telescope Making
Stephen F. Tonkin

Observing Meteors, Comets, Supernovae
and other Transient Phenomena
Neil Bone

Astronomical Equipment for Amateurs
Martin Mobberley

Transit: When Planets Cross the Sun
Michael Maunder and Patrick Moore

Practical Astrophotography
Jeffrey R. Charles

Observing the Moon
Peter T. Wlasuk

Deep-Sky Observing
Steven R. Coe

AstroFAQs
Stephen F. Tonkin

The Deep-Sky Observer's Year
Grant Privett and Paul Parsons

Field Guide to the Deep Sky Objects
Mike Inglis

Choosing and Using a Schmidt-Cassegrain Telescope
Rod Mollise

Astronomy with Small Telescopes
Stephen F. Tonkin (Ed.)

Solar Observing Techniques
Chris Kitchin

Observing the Planets
Peter T. Wlasuk

Light Pollution

Responses and Remedies

Bob Mizon

With 139 Figures
(including 98 in Colour)

Springer

Cover illustrations: The left-hand photograph shows a pristine night sky in the English countryside. The right-hand photograph shows the stars erased by glare from an indifferently lit road junction. The circular insert is an all-sky photo of the night sky, invaded by waste upward light from distant towns. Can you spot comet Hale-Bopp in one of the darker parts of the sky?
Photographs, respectively, by Bob Mizon, John Ball and Mike Tabb.

British Library Cataloguing in Publication Data
Mizon, Bob
Light pollution : responses and remedies. – (Patrick Moore's practical astronomy series)
1. Light pollution 2. Cosmology
I. Title
523.1
ISBN 1852334975

Library of Congress Cataloging-in-Publication Data
Mizon, Bob, 1946-
Light pollution : responses and remedies / Bob Mizon
p. cm. -- (Patrick Moore's practical astronomy series, ISSN 1431-9756)
Includes bibliographical references and index.
ISBN 1-85233-497-5 (alk. paper)
1. Light pollution. 2. Astronomy--Observations. I. Title. II. Series.
QB51.3.L53 M59 2001
522--dc21 2001040008

Patrick Moore's Practical Astronomy Series ISSN 1617-7185
ISBN 1-85233-497-5 Springer-Verlag London Berlin Heidelberg
a member of BertelsmannSpringer Science+Business Media GmbH
http://www.springer.co.uk

Printed in Singapore

Typeset by EXPO Holdings, Malaysia
Printed and bound by Kyodo Printing Co. (S'pore) Pte. Ltd., Singapore
58/3830-543210 Printed on acid-free paper SPIN 10836534

…and we are for the night. Shakespeare: *Antony and Cleopatra*

Acknowledgements

Thanks for all their generous help and encouragement in preparing this book, for the supply of images and for wise advice, go to several members of the British Astronomical Association's Campaign for Dark Skies: Chris Baddiley, John Ball, Graham Bryant, Bill Eaves, John Mason, Ian Phelps and Mike Tabb; also to Patrick Baldrey (ILE/Urbis Lighting), Ninian Boyle (Venturescope), David Coatham (ILE), David Crawford (IDA) and colleagues, Alan Drummond (BAA), Syuzo Isobe, Penny Jewkes, Eric Jones (SSE Museum of Electricity, Christchurch), Ian King, Marlin Lighting Ltd., Nigel Marshall, Richard Murrin, David Nash, John Procktor (D.W. Windsor Ltd.), Alistair Scott (Urbis), Woody Sullivan (University of Washington, Seattle), Nik Szymanek, Martin Morgan Taylor, Steve Tonkin and fellow members of the Wessex Astronomical Society, and Paul Williams for all the photoprocessing. My special thanks to Pam Mizon for her patience and support.

Photographs

All photographs in this book are by the author unless otherwise credited. Every effort has been made to trace owners of attributable material.

Contents

Introduction

A national schools' test paper in science, issued in the early 1990s and aimed at fourteen-year-olds, contained the following question: "If you look at the sky on a clear night, you can see any of the following: galaxies, moons [*sic*], nebulae, planets, stars. Tick the three that are outside the solar system". This question assumed a lot, the questioner seeming blithely unaware of a cruel irony of modern education. We require our children to appreciate the wider universe in the school curriculum, but the vast majority of them see very little of their universe because of the pall of wasted light which hangs over every city – and many villages and rural spaces – in the developed world.

Thousands of stars should be visible to the unaided human eye from a dark place, but it is becoming increasingly difficult to find such places. There are sites in modern town centres where nothing external to the solar system is ever seen in the sky.

The casual observer who draws pleasure from a chance encounter with the starry sky on an out-of-town trip; the amateur astronomer who probes the near and far cosmos from a back garden; and professional scientists, modern Columbuses who use ever more subtle machines and methods to bring the wonders of the far universe down to Earth: all these are increasingly challenged by the adverse impacts of modern technology upon the environment above.

Waste light, radio interference, space debris, aircraft contrails all contribute to, and will aggravate if unchecked, the increasing barrier between the human race and its cradle, the cosmos. We all are made of star-stuff, nearly every atom in our bodies created in some distant and probably long-dead star, some explosive event whose reverberations have long since dissipated; and whatever is left of our material selves, when our planet finally sears in what Bertrand Russell called "the vast death of the solar system", will be redistributed, recycled, into the cosmic depths which we can no

Fig. 0.1. Bob (left) with French dark-sky campaigner Alain Legué, at the European Astronomy Congress, Nantes, France, April 2000 (photo: Shelley Fey).

longer, at the beginning of the twenty-first century, properly see and appreciate.

Michael Crichton, author of *Jurassic Park* and many other works of both fiction and non-fiction, summed up the predicament faced by the observer of the heavens, when he wrote in his book *Travels*:

> The natural world, our traditional source of direct insights, is rapidly disappearing. Modern city-dwellers cannot even see the stars at night. This humbling reminder of Man's place in the greater scheme of things, which human beings formerly saw once every twenty-four hours, is denied them. It's no wonder that people lose their bearings, that they lose track of who they really are, and what their lives are really about.*

Light Pollution: Responses and Remedies is not a "science book" in the usual sense. It is in the *Practical Astronomy* series not only because it offers a selection of objects which may be studied in moderately light-polluted skies, but also because its contents may point to courses of action which astronomers, be they ardent campaigners (Fig. 0.1) or mildly concerned individuals, can follow in order to contribute to the alleviation and eventual solution of the skyglow problem. This book deals with human perceptions as much as with the discipline of astronomy; with our aspirations and needs as much as with our technical achievements. It explores one of the saddest paradoxes of modern life: the fact that our developing technology can provide us with stunning images of the near and far universe, and at the same time blind our eyes to the stars above.

---

* Michael Crichton, *Travels* (Pan/Macmillan, 1988. ISBN 0-330-30126-8)

Chapter 1
Light Old and New

The Sun, Moon and stars have been our lights since the earliest times. We have learned to harness and domesticate fire, and we have much more recently filled our homes and streets with artificial lighting. We are, however, in danger of losing touch with some of those *natural* lights which have guided, comforted and inspired the human race throughout its history.

1.1 The Limits of Human Vision

Observing the starry sky on a clear, moonless night, from a location well away from population centres (Fig. 1.1), can be an experience both mystical and unsettling. The enormous dome of the heavens may engender feelings of insubstantiality and remoteness, and the thousands of stars visible to the unaided eye crowd out half-remembered constellation patterns. When the Milky Way rides high, we marvel, as did Chaldean farmers and the hominids of the Great Rift Valley, at this river of light dividing the sky. The sense of wonder is hardly diminished, and may well be enhanced, by the fact that, unlike our forebears, we know that above us arches not a misty river in the sky, but the galaxy in which we live, on a small planet around an ordinary star.

Estimates of the number of stars in the Milky Way, a medium-sized spiral galaxy like countless others

Fig. 1.1. A clear night sky: Orion looks down upon the unlit village of Ansty, in Dorset.

(Fig. 1.2), vary. It may hold as many as two hundred billion (2×10^{11}) stars, in a matrix of clouds of gas and dust from which future solar systems will arise. For

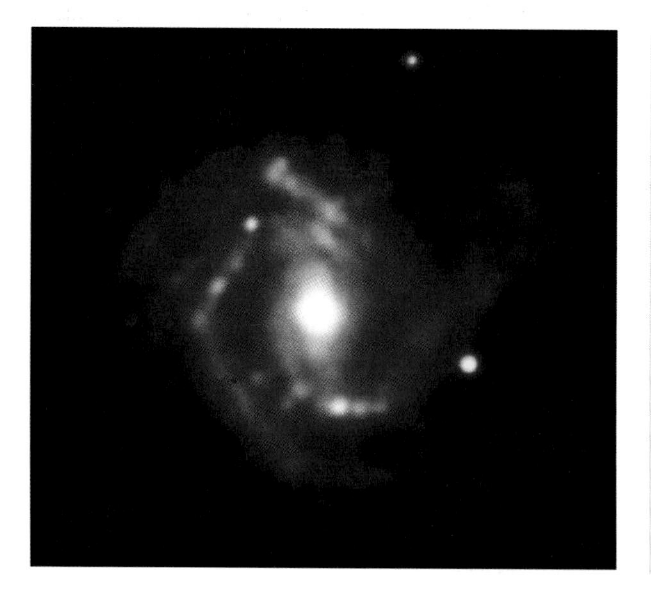

Fig. 1.2. M61: a Milky Way type galaxy (CCD image: David Strange).

every human being on Earth, there may revolve around the galactic centre some thirty-three stars.

From a dark rural retreat, our eyes see between two and three thousand stars only, out of all these teeming drifts; there are people with exceptional vision who can see perhaps seven thousand. The human eye, like all mammalian eyes, is a complex organ: the light-sensitive cells (rods and cones) of the retina at the back of the eye, the optic nerve, transporting impulses from the retina to the brain, and the various structures (lens and muscles) involved in focusing incoming light onto the retina, all combine to convert photons into electrical impulses which the brain can further convert into a sharp image.

Like a camera, the eye focuses light from the environment onto a sensitive surface, controlling the amount of incoming light, and creates a pattern perceived as an image of that environment.

The retina is an extension of the forebrain. Its light-sensitive rods and cones, fairly similar in structure though the rods are more cylindrical, are embedded in the "photoreceptor layer" of the retina. In the human eye, rods outnumber cones by about twenty to one: each eye has about six million cones, closely packed into the centrally located fovea, where most light is focused. Cones contain three visual pigments, differentiate colours, and work mainly in daylight. The 120 million rods, with only one photosensitive pigment, have weaker visual acuity and are found more loosely distributed right across the retina. Rods are collectively more sensitive than cones, and largely take over visual function in the dark. At night things appear to lose colour, as rods dominate. We can just about see colours in bright stars, but optical aid is usually necessary really to appreciate this. The clustering of cones and absence of rods in the fovea explains why averted vision works, and why faint objects in a telescope field may suddenly appear if we look slightly to one side of them: paradoxically, we see objects more clearly at night by not looking straight at them, allowing more rods to function. The light-sensitive pigment in rods, rhodopsin (visual purple), is a complex molecule. It is formed by a reversible chemical reaction in low-light conditions, and the resultant process of dark adaptation can take up to 30 minutes in total darkness. The use of a red light at night is recommended for observers, because visual purple will not react to its particular wavelengths and adapted night vision will not be affected.

Fig. 1.3. A 30 s exposure: Jupiter and Saturn in Taurus, 2000 Aug 11.

Photographs taken with a relatively short exposure (short in astrophotographers' terms) of about thirty seconds (Fig. 1.3), using a fixed camera and a fast film (for example, ISO400) reveal many times more stars than we can ever appreciate with the visual equipment Nature has allotted to us.

Our limited vision is only one of the natural factors which combine to rob us, even in the darkest places, of the vast majority of heaven's night-time lights. Our next big problem if we want to see stars is that we have to observe the rest of the universe through a multilayered and busy sea of gases, the Earth's atmosphere (Fig. 1.4), with a total mass of the order of 5×10^{15} tonnes, about one-millionth of the mass of the Earth as a whole.

The lowest level of the atmosphere, the troposphere, is deepest at the equator (around 28 km/17 miles) and at its most shallow at the poles (around 7 km/4.5 miles). In this layer most of the world's weather occurs, and life processes are maintained. Temperature decreases in the troposphere with increasing altitude, and great air masses slowly turn and interact as tem-

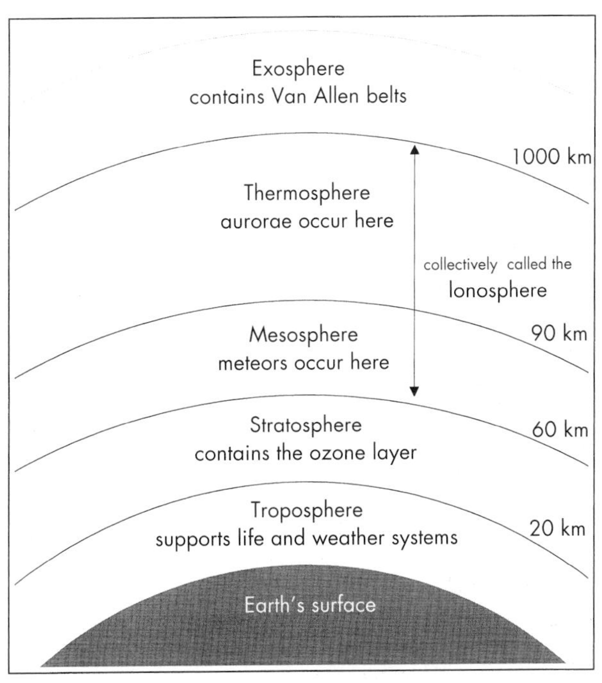

Fig. 1.4. The atmosphere (diagram courtesy of Nigel Marshall).

perature differentials and the planet's rotation and relief steer them. What we normally understand by our "atmosphere", with its active appearance and constantly changing weather, is merely the lowest few kilometres of a much vaster entity.

Above the troposphere extends the thermally more stable but much more tenuous stratosphere, devoid of clouds, with a ceiling at about 60 km/37 miles. High-flying modern jet aircraft pass through the base of the stratosphere, but only rockets have ever climbed into its higher reaches. Next come the mesosphere (lower ionosphere), whose ceiling is at approximately 90 km/56 miles, and the thermosphere (upper ionosphere), extending far upwards to about 1000 km/625 miles. Cosmic dust particles, swept up by the Earth in its 100 000 km/h/66 000 mph progress around the Sun, burn brightly in the mesosphere as they encounter layers dense enough to abrade them by friction, and meteors are the visible result. The upper stratosphere and lower mesosphere contain the ozone layer, which plays a vital part in shielding us from the Sun's lethal ultraviolet radiation.

The ionosphere also contributes to our well-being, by soaking up incoming X-rays. Here, rarefied gases are ionised as they absorb solar radiation, enabling high-frequency radio transmissions to be sent around the world, as signals are reflected back downwards from the ionised layer. Radio waves of similar wavelengths from space are scattered away from the planet by the ionosphere. The beautiful polar aurorae, described in more detail later, also inhabit this region. Further upwards still, the exosphere, an assemblage of loose particles rather than a layer, and most famously containing the Van Allen radiation belts, gradually fades into the near-vacuum of outer space.

Confined as we are to the bed of this turbulent ocean of air, it is perhaps surprising that we see as much as we do of what lies outside the Earth's atmosphere. Its ceaseless motions cause the stars apparently to twinkle (scintillation), as their light traverses the last disrupting millisecond of its enormously long journey to our eyes. Astronomers use the term "seeing" to describe the relative steadiness and transparency of the atmosphere, as judged by the appearance and behaviour of a telescopic image. The old but reliable Antoniadi scale is still in use by many observers when recording their observations, its criteria being:

- I perfect seeing, without a quiver;
- II slight undulations, with appreciable intervals of calm;
- III moderate, with large tremors;
- IV poor, with constant undulation;
- V very bad seeing, scarcely allowing the making of a rough sketch.

Ever-present droplets and minute particles (aerosols), suspended in countless numbers in the atmosphere, interfere by absorption and scattering with much of the light passing through, causing extinction and distortion of objects near the horizon. There are many different kinds of aerosols. They include mineral particles lifted from deserts, volcanic ash, salt crystals evaporated from sea spray, pollen grains, bacteria, spores, and minute waste products of our industrial society. Aerosols play an important rôle as condensation nuclei for atmospheric water droplets.

Water suspended in the atmosphere (Fig. 1.5) is the chief culprit in attenuating or completely hiding the light from distant bodies, and not only on cloudy nights. Many modern optical astronomers endure the discomfort of high-altitude spells in remote mountaintop observatories, sited above the worst of the turbu-

Fig. 1.5. Clouds: the astronomer's *bête noire*.

lence and cloud, beneath clearer air through which the night sky is more effectively observed. A modern manifestation involving water in suspension is beginning to occupy the minds of astronomers, especially in Europe and North America, as they see thin condensation veils, left by merging aircraft contrails, spreading slowly across the skies during even the clearest day (Figs. 1.6 and 1.7). The increase in air traffic generally means that this will possibly be the next problem that the astronomical community will have to take on board in its constant battle for a clear view.

Those of us living in brightly lit towns, and seeking the best possible experience of the star-strewn heavens without having to climb a mountain, may yet escape to our own preferred dark place, in order to see fainter night-sky objects such as the Andromeda spiral (M31), the nearest major galaxy to our own, at a distance of 2.2 million light years (Fig. 1.8). M31 may contain as many as 250 billion (2.5×10^{11}) stars. The light from this dim, oval glow, travelling towards us at 300 000 kilometres/186 000 miles per second through a foreground of Milky Way stars, comes from the furthest thing that we can perceive directly with any of our senses. To say that a dark sky lets us see further is most certainly an understatement.

Even the darkest skies, however, are not as black as we might imagine. We can never really see a totally

Fig. 1.6. Drifting contrails draw a veil across a clear sky.

light-free backdrop to the scattered stars, since the night sky has a natural "glow". This faint luminescence is caused by sunlight reflected from myriads of dust particles distributed throughout the solar system, and by impacts of energetic particles from the depths of the universe, and also from the Sun, upon the uppermost layers of the Earth's atmosphere. A tiny contribution comes also from large numbers of Milky Way stars, and even from remote galaxies, which are individually beyond naked-eye visibility, but combine to add to the general dimly luminous effect.

In spite of all the natural factors seemingly conspiring to draw a veil across the stars, our poor view of the heavens has played a continuous and enormously important part in human affairs. The fact that *Homo sapiens* is the only terrestrial mammal equipped with the musculature to look upwards whilst maintaining balance for sustained periods may have been the key to the door leading to our faculty of wonder, and to our contemplation of a "scheme of things" greater than our immediate surroundings. The explorer, the mystic, the poet and the scientist residing to differing degrees in all of us may largely owe their presence to the starry vault which early hominids tried somehow to understand, ached to relate with, during the nights of millions of years ago (Fig. 1.9).

Fig. 1.7. A veil of contrails over Northern Europe (courtesy Deutsche Forschungsanstalt für Luft- und Raumfahrt).

Fig. 1.8. Two-million-year old light from M31 (photo: Alan Jefferis).

Fig. 1.9. Sky-wonder: a god (Jupiter) and a brighter goddess (Venus) meet (1988 February 29).

Much of the mythology of modern cultures and religions reflects this sky-wonder, and any primary school teacher will tell you that there are two things in the early science curriculum which light up the eyes of modern five-year-olds: dinosaurs and outer space, both of them remote, magical, vast, tempting with the twin lures of the totally unattainable and the visually splendid.

During the second half of the twentieth century, public interest in the cosmos grew as modern space missions, from *Sputnik* through *Voyager* and *Apollo* to *Galileo* and *Cassini*, vastly increased our knowledge of planetscapes and processes within the solar system. Ever more sophisticated radio dishes, earthbound detectors and orbiting telescopes continue to reveal undreamed-of facets of the deeper universe. However, the opportunity to experience the night sky *directly* remains important. The assumption that the Earth is all that exists, and that stars and planets belong in picture books, on cinema and TV screens and computer monitors, because they cannot be seen in skies invaded by wasted upward light, is an ultimate and dangerous vanity.

The Range of Natural Radiations

The complete electromagnetic spectrum of radiation (Fig. 1.10) rains down incessantly upon our planet: the whole range of cosmic wavelengths, from those of the very longest radio waves (10^5 m) to the piercing beams of the most energetic X-rays (10^{-11} m) and gamma rays (10^{-14} m), floods into the topmost layers of the atmosphere. Also, high-velocity particles, or cosmic rays, penetrate the atmosphere to various depths, and countless trillions of ghostly neutrinos pass unhindered right through the planet, and everything on it, every day. The fact that our atmosphere is transparent only to certain radio and visible-light radiations, the so-called radio and light "windows", means that, although we are protected from the worst that high-energy radiation might do to our bodies, we have to send spacecraft up above the atmosphere in order properly to study the universe at those wavelengths denied to us by our shield of air.

Today's high-technology orbiting observatories and terrestrial radio telescopes will doubtless be superseded in the far future by remotely controlled instruments working through the frigid, two-week long nights on the far side of the Moon. Although set on the Moon's dusty surface, these detectors will be literally in outer space, with an unspoiled view of the cosmos. If they are located at the centre of the lunar far side, they will be shielded by nearly 3500 kilometres/2200 miles of rock from Earth's radio "noise", and from earthlight or earthshine, the lunar equivalent of moonlight upon the Earth (Fig. 1.11). The Earthward side of the Moon is

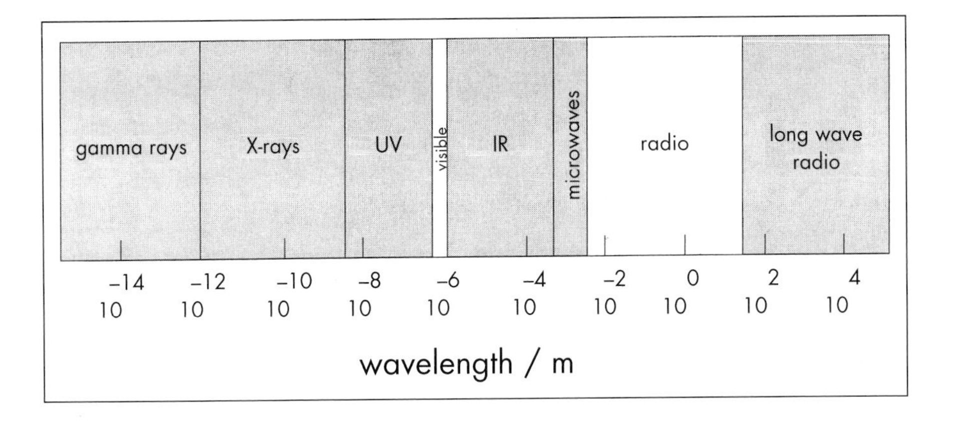

Fig. 1.10. The electromagnetic spectrum (diagram courtesy of Nigel Marshall).

Fig. 1.11. Earthshine.

bathed at night in sunlight reflected from the bluish planet, four times wider in the Moon's sky than the Moon in ours. The Earth is never seen from about 41% of the Moon's surface.

Throughout scientific history, and especially as telescopes have increased in power since the early seventeenth century, students of the night sky have announced the existence of ever greater numbers of luminous objects, showing a wide range of forms and activity. Galileo Galilei, observing through a fairly primitive refractor in 1610, realised and reported that "the Milky Way is nothing but a mass of innumerable stars planted in clusters", and today's monster telescopes continue to fish up ever more distant and fainter galaxies. Increasingly, since Galileo turned his spyglass to the Milky Way, it has occurred to many people that, if the Earth's atmosphere is transparent to the wavelengths of visible light, and if there are so many stars and galaxies in whatever direction we look, then the entire dome of heaven ought to be covered with luminous points, and the night sky should be bright instead of dark.

So why is the sky dark at night? This conundrum is known to science as Olbers' paradox, after Heinrich Olbers (1758–1840), a German physician and amateur astronomer who was the first to discuss the problem in depth, in the 1820s. Astronomers remember Olbers nowadays more for his treatment of this perplexing question than for his equally meritorious work on comets and their orbits, and his discovery, from his own observatory in Bremen, of the asteroids Pallas and Vesta, the second and fourth minor planets to be detected.

Olbers' paradox is resolved chiefly by the fact of the expansion of the universe. The unchanging, uniform and infinite cosmos of bygone centuries is no more, and we realise that the attenuation and reddening of the energy of very distant starlight, caused by the enormous velocities of recession of the fleeing galaxies, explain why we are not dazzled by those distant stars as well as by the Sun. So the darkness of the night sky seems chiefly due to what E.R. Harrison of the University of Massachusetts eloquently called "the infrared gloom of the Big Bang". With clouds of silicon and carbon dust (Fig. 1.12) within

Fig. 1.12. Dust clouds are prominent in this photograph of the Milky Way. The brightest object is Jupiter.

our own galaxy also contributing to the absorption and dimming of starlight, we now have to accept that it is not just our turbulent atmosphere and the wasted energy from the lights of modern civilisation which deny us a deep view of the stars, but the history and nature of the universe itself.

Sunlight

Our G-type, subdwarf star (meaning, confusingly, that it is classified as a little bigger, not smaller, than a dwarf star) drives all our weather, grows our food, and, through fossil fuel energy, powers our vehicles, lights our homes, and runs our machines. For all the Sun's apparent violence, it is a reassuringly stable star (Fig. 1.13). Life on this planet would be impossible if there were a variation of just a few per cent in its energy output, the product of the conversion into helium of some 564 million tons of hydrogen every second, of which four million tons are annihilated and radiated as energy.

Our ration of this energy is but a tiny fraction of the whole, the mean daily value of the solar radiation striking the top of Earth's atmosphere being 342 W/m^2. A simple way to appreciate the nature of solar light is to

Fig. 1.13. A stable star: sunset at Slingsby, in Yorkshire.

Fig. 1.14. The solar spectrum begins its daily journey across the kitchen wall.

place a prism on a Sun-facing windowsill, and watch the stack of vivid colours projected onto the opposite wall move slowly across the room as the hours pass – evidence of the rotation of the Earth seen from your armchair (Fig. 1.14).

The Sun lights the night sky indirectly in the form of the *aurora borealis* (in the southern hemisphere, the *aurora australis*). High-speed electrons ejected in clouds from the Sun are funnelled along the Earth's magnetic field lines towards the poles. They interact with atoms and molecules of oxygen and nitrogen in the air, energising them with consequent emission of red and green light in ghostly arcs, ribbons and patches, moving, sometimes remarkably rapidly, across the sky. Vigorous "storm" aurorae sometimes spill down into lower latitudes. For example, there were intense all-sky displays visible from all over Europe and North America on 1989 March 13–14 and 2000 April 6–7 (Fig. 1.15), reportedly disabling satellites and affecting power supplies.

The planets, borrowing their lustre from the Sun, cast a steadier light than stars as they weave their slow paths through the background patterns of the constellations of the Zodiac. With a little practice, it is not difficult to tell which planet is being observed.

Mercury's bright dot, comparable to the brightest stars, is rarely seen by the casual observer, and then

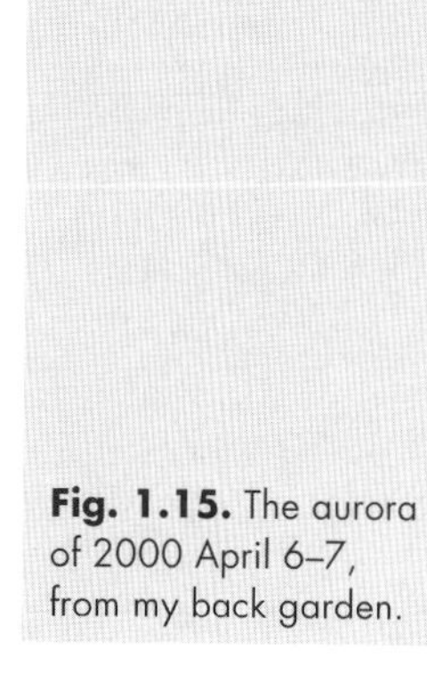

Fig. 1.15. The aurora of 2000 April 6–7, from my back garden.

usually between trees or housetops (Fig. 1.16), as it never moves further than 28° away from the Sun as seen from Earth. It is therefore seen in a twilit rather than a really dark sky, and, at dawn, will be lost to the brightening eastern glow, while at dusk, it will chase the Sun down and soon disappear behind horizon objects.

Venus' highly reflective sulphuric acid clouds cause it to appear brilliantly white, and this "morning star" or "evening star" (Fig. 1.17), never more than 46° away from the Sun, can appear so uncommonly bright (up to a glittering magnitude –4.7) that it is often reported by non-astronomers as a mysterious UFO in the sky, espe-

Fig. 1.16. Mercury sets below the Pleiades, 1996 April 24.

cially if seen from a moving vehicle, as its great distance causes it apparently to "follow" the observer. Try an interesting experiment: when Venus is at its most brilliant, can you find an observing site dark enough to be able to verify whether Venus really can cast shadows on a light-coloured surface, as many astronomers have claimed?

The iron-rich and rusty soils of Mars give it its reddish appearance, and its identification with the god of war may stem from the association of Mars' colour with that of blood, fire and tearful eyes. On its journey around the ecliptic zone, Mars occasionally passes "near" the red giant stars Aldebaran (magnitude 0.85), and Antares (magnitude 0.96), and the planet may well be confused with either of them (Fig. 1.18). Indeed, Antares means, in Greek, "rival of Mars", but the planet may be distinguished from these stars by its steadier light.

The giant gas planets Jupiter (Fig. 1.19) and Saturn are visible for most of the year. Older and wiser gods, they have a more sallow, yellowish look, but they can be very bright. Jupiter may attain magnitude –2.3, and Saturn –0.3. Many observers see Saturn as somewhat dull in hue compared to its bigger and brighter neighbour Jupiter.

Fig. 1.17. The evening star: Venus at twilight, 1996 October 15.

Moonlight

To demonstrate that there is no such thing as true moonlight, but merely reflected sunlight, simply take a long-exposure photograph (30 seconds to one minute)

Fig. 1.18. Mars (below centre) nears its red rival Antares (bottom left), dawn, 2001 February 24.

Fig. 1.19. Jupiter and Bob's 21-cm/8.5-inch reflector.

of a landscape with a tripod-mounted camera on a clear night, when the Moon is up, and is full or nearly so: the resulting photograph (Fig. 1.20) will be indistinguishable from a sunlit, daytime scene, with green grass and a blue sky, though the presence of stars in the sky will give the game away.

The amount of sunlight reflected away by the Moon is small compared with what falls on it. The albedo of the Moon is 0.7 on a scale of 0 to 10, representing a reflective capability of only 7%. Even this poor ration of night-time light was enough, however, to allow human activity to continue after sunset, beneath the Moon, before the advent of artificial lighting.

As the Moon revolves about the Earth, we can follow the progress of night and day on our satellite as it goes through its cycle of phases. It is surprising how many people still believe that the Moon has a permanently dark side, in spite of the fact that they can see the frigid lunar night pass slowly across the Moon's Earthward face over a period of four weeks (Fig. 1.21).

Fig. 1.20. Moonlight: a midnight Moon and a 20-second exposure create a daylight scene.

Our ancestors venerated the Moon for a good reason: since the dawn of the human adventure on Earth, it has provided light, for at least some of the time, at night. The presence of the Moon at night might have made the difference between a hunter-gatherer community thriving or starving in prehistoric times, and in the twenty-first century there are still many groups of people, remote from the world's artificial infrastructure, who depend upon this light at night rather than that provided by a power company. The human eye has evolved to take advantage of this low-level night-time illumination, and can adapt itself to surprisingly dim conditions in a very short time, if given the chance. The hormone rhodopsin ("visual purple") stimulates the cylindrical rod cells of the retina in low-light conditions, and, if we have been exposed to bright light, we can usually achieve effective dark-adaptation beneath the night sky in an unlit location within a few minutes, though sensitivity may continue to increase, especially in younger eyes, for up to half an hour.

Many modern humans, when going outside at night from their brightly lit houses, may glance at the night sky and, without giving their eyes a chance to adapt, too readily conclude that there is little to see up there even if it is reasonably dark outside. In all the millennia

Fig. 1.21. Day and night on our satellite. A five-day-old Moon, with prominent Earthshine, occults the star ZC1070, 1990 April 29.

before the invention of bright electric lighting, the stars must have been a striking and familiar sight to people venturing out from less generously lit interiors.

Starlight

As a fledgling teenage stargazer in the early 1960s, proud of my cheap 3-inch refractor and voraciously reading and rereading Patrick Moore's books, *Norton's Star Atlas* and the *Larousse Encyclopaedia of Astronomy*, I was constantly surprised by the stars. I saw plenty of them at that time (Fig. 1.22) from the small Shropshire town in which I then lived.

My first surprise was that the nearest star to the Sun, Proxima, turned out to be invisible, not only because it is more than 62° south of the Celestial Equator while I lived at latitude 52° north, but also because it is a dim red dwarf massively upstaged by its neighbour α Centauri. Yet one of the brightest of the summer stars, the blue-white beacon Deneb (Fig. 1.23), at magnitude +1.3, lay more than 1600 light years away! The Romans

Fig. 1.22. The Milky Way flows through Cassiopeia and Cygnus in a dark, rural night sky.

were still marching along Watling Street,* their main road from Dover through London to North Wales, which passed within a few miles of my boyhood home, when Deneb's light began its journey to my younger eyes. I was similarly surprised to find that the famous Pole Star (Fig. 1.24) is only the forty-ninth brightest in the sky, at magnitude +1.9.

This concept of stellar magnitude harks back to classical times, when the brightest stars were assumed to be the biggest (Latin *magnitudo*, size, bigness). The scale of magnitudes has its origins with the Greek astronomer Hipparchus, whose catalogue of stars (*c.*140 BC) is the first known to divide them into orders of brightness, using terms like "bright" and "small". Three centuries later, Ptolemy declared in his *Almagest* (*c.* AD 140), which often recalls Hipparchus' ideas, that the brightest stars would be of the first magnitude, prominent but less "important" ones of the second, and so on. The faintest stars visible were to be of the sixth magnitude. All this can be rather subjective, and hardly lends itself to scientific accord and accuracy, so

*According to R.H. Allen (*Star Names*, 1899), the name *Watling Street* is derived from the Anglo-Saxons' *Waetlinga Straet* ("Giants' Way"). It was one of their names for the Milky Way.

Fig. 1.23. Deneb is the brightest star in this 3-minute exposure from Child Okeford, Dorset.

nowadays it is agreed that a star of magnitude +6.0 will be one hundred times fainter than a star of magnitude +1.0. It follows that a star of, for example, magnitude +5.0 is 2.512 times brighter (2.512 being the fifth root of 100) than a star of magnitude +6.0, and so on up and down the scale. The plus sign is important, since the magnitude of the brightest star in the night sky, Sirius, becomes by this reckoning −1.4. Still not the simplest of schemes, it serves astronomers well enough for them to hold on to it.

From a very dark place, some keen-sighted observers claim to see stars even fainter than the sixth magnitude, down to the seventh. A modest telescope, for example a 75-mm refractor, will show stars down to about eleventh magnitude in a dark sky.

The light which we receive from the stars is by no means negligible: it is about one-fifteenth of that which we receive from a full Moon. Starlight is the result of nuclear transformations deep within stellar cores, and in their long lifetimes, typically billions of years, the stars create new elements from other, less complex ones. For example, a solar-type star will create most of its energy from the continuous crushing together in the 17-million degree furnace of its core of vast numbers of hydrogen nuclei to make helium. Four hydrogen nuclei become one helium nucleus, a process which involves mass loss and the liberation of energy. As a result of

Fig. 1.24. The Pointers of the Plough indicate the Pole Star, high above the marquee of a school's summer camp.

this relentless manufacture of helium, the Sun annihilates millions of tons of its mass every second, to be dispatched into space as energy in its various forms of electromagnetic radiation.

Many of the other elements so familiar and vital to us on this planet, for example oxygen, silicon and carbon, were made by stars a little more massive than the Sun. The comparative scarcity, and resultant high price, of heavy elements such as gold – the International Gold Corporation estimates that all the gold in the world, mined and unmined, could be formed into an 18-metre cube – stem from the fact that they are created during the final paroxysm of those rare stellar blast-furnaces, supernovae (Fig. 1.25). These exploding stars, ending their cosmically brief lives by rapidly ejecting nearly all their mass outwards into space, are the only ones hot enough to forge the heaviest of elements. Since there is, on average, only one supernova outburst every century in any one galaxy of a hundred billion stars, gold is in short supply in the universe.

The most recent supernova visible to the unaided human eye was first seen on 24 February 1987, about 160 000 years after its eruption in the Large Magellanic Cloud (LMC), by Ian Shelton at the observatory of Las Campanas in Chile. The once lowly tenth-magnitude star Sanduleak –69 202, suddenly the most famous

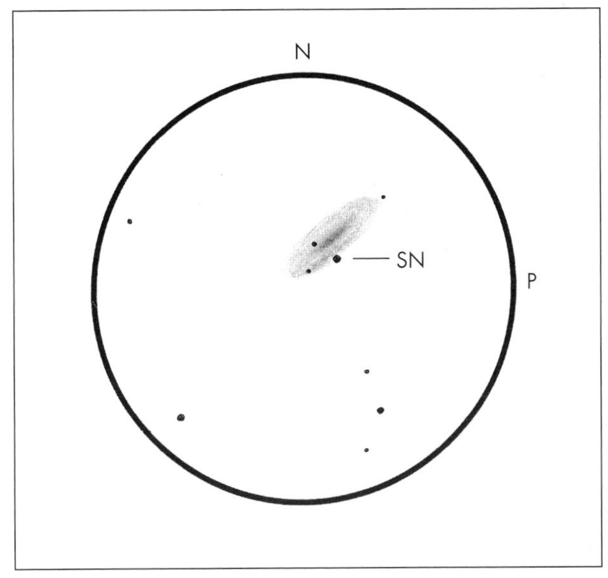

Fig. 1.25. Supernova 1993J in M81 (NGC 3031).

object in the sky, eventually peaked as SN1987A at an apparent magnitude of +3. The last naked-eye supernova in our Galaxy was recorded by Danish astronomer Tycho Brahe in 1604. We are well overdue for another. Occasionally, an unexpected nova (Figs. 1.26a/b) reaches naked-eye visibility. The nova will not have the *éclat* of a supernova, but may be bright enough for its progress to be monitored with the naked eye or with modest optical aid for a few days or weeks, finally disappearing again from view. In spite of their name, novae are old stars rather than new ones, members of binary systems in which a white dwarf star has material "dumped" upon it by a larger companion.

A clear view of the stars led, in ancient times, to their being grouped into constellations, many of which survive to this day as a convenient naming and reference system. Five thousand years ago, the Sumerians and the Egyptians had already established a tracery of creatures, symbols and heroes in the sky (Fig. 1.27), and two thousand years ago, Ptolemy listed forty-eight constellations from former times. Today, eighty-eight constellations mark out the heavens, and on a clear night, with a simple chart and a good imagination, we can try to retrace the thoughts of the ancients as they found their way around the night sky, which served them as clock, compass, calendar and oracle, and as a scientific and religious primer. The constellations have kept their identities over the millennia, and their

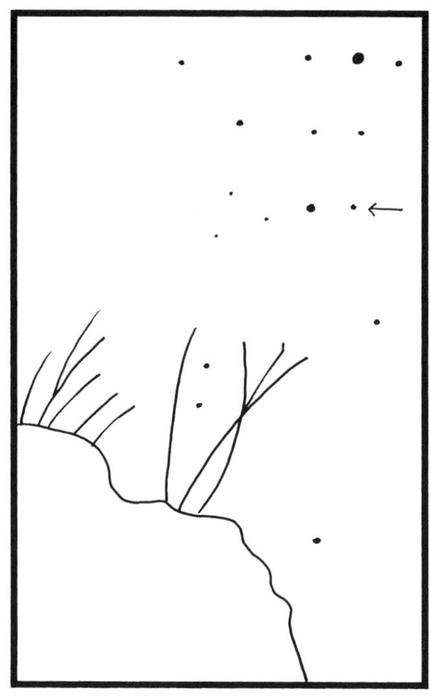

number has gradually increased, facts which bear witness to the continuing fascination which the sight of the starry sky exerts over humans, and to our need to observe and interpret what we see in the environment above us.

Fig. 1.26. Nova Aquilae 1999 December 4. The "new" star is near Delta Aquilae, and is arrowed on the accompanying chart. Altair is the brightest star at the top of the photo.

The Zodiacal Light

A common sight in a starry, moonless sky as seen from the countryside, before wasted light from distant cities and local lamps began to intrude, was the Zodiacal Light, a faintly glowing cone, slanting upwards from its base along the horizon, seen to the east shortly before dawn or to the west shortly after dusk (Fig. 1.28). Brighter than parts of the Milky Way, it is best seen when the ecliptic, along which its axis lies, stands nearly vertical to the horizon in spring or autumn.

This light is due to scattering of sunlight by vast numbers of dust grains which litter the solar system, many left behind by visiting comets. Giovanni Cassini was the first to realise this in 1683. The particles involved are mostly between 0.1 and 0.2 microns in

Fig. 1.27. An ancient constellation: Orion, the Osiris of the Pharaonic Egyptians.

diameter, and tend to be concentrated in a lens-shaped cloud, centred on the Sun, and aligned with the plane of the planets' orbits. This explains why the Zodiacal Light stretches along the ecliptic.

All this cosmic light which falls from the moonless dome of stars may not necessarily be bright, though to a dark-adapted eye it can be surprisingly effective in revealing one's surroundings: the French playwright Corneille testified to this in *Le Cid*, written in 1637, long before the era of all-night artificial lighting. In the fourth act of the play, Don Rodrigo (the Cid) spies the approaching Moorish battle fleet by starlight, "*cette obscure clarté qui tombe des étoiles*" ("that faint light that falls from the stars").

Today, at the beginning of the twenty-first century, the night's natural light, emanating from a variety of sources and even indirectly from the Sun, has been erased and veiled as never before in human history (Fig. 1.29). The last fifty years have seen the stars go out for much of the world's population. It does not require much investigation to know why.

Fig. 1.28. The Zodiacal Light from La Palma (photo: Alan Drummond).

1.2 Light Pollution: the Problem Defined

For three million years, on every clear night, human beings have been able to behold the spectacle of the starry heavens, spangled with thousands of stars and traversed by the Milky Way. Nature's grandest free show has spurred us to consider our place in the great scheme, has given rise to many themes and aspects of our cultures and religions, and has inspired both artistic achievement and scientific endeavour.

The unspoiled starry sky is, unofficially but undeniably, a site of special scientific interest, and an area of outstanding natural beauty. Now, ill-directed artificial light is quietly stealing it away from most people in the developed world. What is the real nature of light pollution? Does it impact only on astronomers? How can it be quantified?

Skyglow

Skyglow is light which is being carelessly, or sometimes deliberately, projected from the ground or a structure, colouring the night sky and reducing the visibility of

Fig. 1.29. Veil across the heavens – light pollution blots out the southern stars.

astronomical objects. It should not be confused with airglow, a word sometimes used for the very faint luminescence ever-present in the night sky due to the Sun's radiation reflected from, or energising, particles both within and beyond the atmosphere.

The cause of skyglow is nowadays well known: upward light from poorly designed lamps (Fig. 1.30), scattered and reflected by aerosols in the atmosphere, returns to the eyes of Earthbound observers, and in areas where this is permitted to occur, some or all of the detail of the night sky may be lost. The effect is not always localised, as the glow above a distant large town (Fig. 1.31) will colour the sky for an observer tens of kilometres away. The light in question can nearly always be classified as wasted, since it is almost never

Fig. 1.30. The massed and mostly poorly directed lights of Canford Heath, Dorset.

the deliberate intention of the owner of the light source to illuminate outside the premises in which the source is located.

In regions or countries where population centres are fairly close together, for example along the Eastern seaboard of the USA, in Belgium, Holland and the United Kingdom, it is possible to travel at night for long distances in nominally "rural" areas without ever escaping strong skyglow from chains of towns and large villages nearby. The skeins of waste light are obvious on satellite images of the Earth at night (Fig. 1.32). Even the smallest village may have a sports facility which floodlights the surrounding area to a great distance, and isolated farms and cottages may be lit with lamps which spread their glare across a far greater area than intended (Fig. 1.33). Even in areas where it is possible to be hundreds of kilometres away from towns, such as in the outback of Australia, cones of light from distant conurbations still taint the sky on the horizon. On a very local scale, a poorly aimed domestic security light with a typically excessive wattage of 300 W to 500 W (what journalist and broadcaster Libby Purves once christened "the Rottweiler light"), will make observation of the night sky difficult, if not impossible, for an observer tens or sometimes hundreds of metres away (Fig. 1.34). Add to this the inappropriate use of very bright lights to illuminate rel-

Fig. 1.31. Skyglow over Poole: the "hot spot" is caused by the floodlights of the Cross-Channel Ferry Terminal.

atively small areas, and the fact that lights are often left on when there can be nobody around who might conceivably need the light or appreciate its effects, and the extent of the problem becomes apparent.

Facing up to the Problem

In 1988, in the USA, the International Dark-Sky Association (IDA), whose founder members include David Crawford and Tim Hunter (Fig. 1.35), took up arms against skyglow. The IDA now has several thousand members in many countries. Soon afterwards, in 1989, the council of the British Astronomical Association, the UK's largest body representing the interests of all who wish to observe and enjoy the sky at night, discussed and authorised the setting up of its Campaign for Dark Skies (CfDS, Fig. 1.36) under the initial chairmanship of Ron Arbour. The time had come, it was felt, to call a halt to the rising tide of waste light in the night sky, and these organisations set themselves the daunting task of not just halting but reversing that tide. Since that time, similar movements have been founded in many countries, and details of these may be found in Appendix 1 at the end of this book.

Fig. 1.32. Europe by Night (Copyright 1996 W.T. Sullivan and Hansen Planetarium).

In 1991 the CfDS carried out a nation-wide survey of over 200 astronomical groups. This included a questionnaire to be distributed to group members, asking for details of their location, and of the visibility of the night sky from it. Respondents were well scattered, some observing from great cities, and others from small towns, villages or isolated rural locations throughout the country.

Of 805 observers, from the casual to the assiduous, who responded, 727 (90.3%) stated that skyglow was visible to some extent in their night sky at home. The great majority (701) of these 727 "positive" respondents commented on the degree of severity of the effect of skyglow on astronomical observations. 211 described it as "noticeable", 453 as "strong", and 37 reported having given up observing the night sky altogether because of "impenetrable" skyglow. The conclusion drawn by the CfDS was that more than 90% of people who wish to see the night sky in the UK, and they are certainly *not* all amateur astronomers, proba-

Fig. 1.33. A farm light shines into a neighbouring astronomer's garden (photo: Graham Bate).

bly suffer light pollution at least noticeable enough to hinder observation. Well over half of these would-be observers have to contend with considerable skyglow. Further details of the survey may be found in the *Newsletter of the BAA Campaign for Dark Skies*, spring 1992, page 2. Collaboration between the CfDS and the National Remote Sensing Centre led to the publishing in 1991 of the now well-known satellite image of a cloud-free UK by night, revealing the extent of light pollution across the country. Similar images of many areas of the world (Fig. 1.37) have been produced by the team led by Professor Woody Sullivan, of the Astronomy Department of the University of Washington (Seattle WA).

How many stars does skyglow take away from us? A pristine dark sky, as seen by an exceptionally keen-sighted young observer, may offer as many as seven thousand naked-eye stars down to magnitude 7 (see "Starlight"). Table 1.1 shows the dramatic reduction in

Table 1.1. Numbers of stars seen at various limiting magnitudes (Source: IDA)

Limiting magnitude (faintest star visible)	Number of stars seen
+7	~7000
+6	~2500
+5	~800
+4	<250 (Milky Way no longer visible)
+3	<50
+2	<25

the numbers of naked-eye stars as they are lost, magnitude by magnitude, to wasted light.

The term "light pollution" has been extended to cover terrestrial light intrusion problems too, but in its astronomical sense it is the veiling effect upon celestial objects of light emitted with an upward component from local or distant luminaires (lamp assemblies). This wasted light does not have to be emitted vertically to cause skyglow. In fact, light traversing a path at a shallow angle above the horizontal from a luminaire will cause more skyglow, since it will encounter more particles and droplets from which to be scattered on its way through the atmosphere.

Fig. 1.34. Warrington, Cheshire: a car lot's security floodlight intrudes into premises well outside its perimeter (photo: Ian Phelps).

Fig. 1.35. Tim Hunter (left) and David Crawford, founders of IDA (courtesy IDA).

The inescapable irony behind light pollution is that, while trying to ensure a more welcoming and easier outdoor night-time environment through lighting technology, we often lose sight of the near and far

Fig. 1.36. The CfDS committee, 1993. Top to bottom: Chris Baddiley, Stuart Hawkins, Paul Kemp, Roy Henderson, David Crawford (visiting from IDA), Pete Welland, John Mason, Graham Bryant, Bob Mizon, Ron Arbour.

Fig. 1.37. The lights of the world by night, from space (Copyright 1994 W.T. Sullivan and Hansen Planetarium).

universe which is being revealed to us in more and more detail almost daily by technologies of another kind, those of spacecraft and imaging. A kind of virtual reality mask is replacing our real experience of that half of our environment stretching above the horizon.

One of the things that nearly everybody "knows" about the Earth as seen from space is that, if an alien craft ever approached this planet, the first sign to the crew of our tenure on it would be a view of the Great Wall of China, commonly said to be the only artefact of humankind visible from outer space. In fact, the Wall is not seen from above our atmosphere, as astronauts have confirmed, being a surprisingly narrow and mostly ruined structure, not strikingly different in colour from its surroundings. The wakes of ships, plumes of steam from power stations (Fig. 1.38), and similar extended objects contrasting strongly with darker backgrounds, are more likely to be seen from above the atmosphere than the Wall. However, as our hypothetical visitors rounded the night side of the Earth, what they *would* see, spread across the darkness, would be the chains and patches of light, by no means all reflected from the ground, thrown up from our towns and cities, road networks, sports and industrial installations, and countless other sites. Then they would know for certain that the planet is inhabited by technologically minded beings, though the energy that we so visibly waste would be likely to count against us in the minds of these advanced observers.

Fig. 1.38. The diffusion of power station and factory steam plumes is striking in this Landsat image of the north-west Midlands of England, taken from an altitude of about 900 km (Copyright Focal Point A-V, Portsmouth).

1.3 Lights and More Lights

The Rise of Artificial Lighting

Since our ancestors crouched around the glow of their wood fires and tallow lamps tens of thousands of years ago, man-made light has been appreciated as a friend to humankind. It allowed its creators to see and perhaps frighten off the approaching predator, to go about their domestic business after sunset, to paint images of great power and beauty in the silent fastness of caves, and to use the night as well as the day for sharing their thoughts, plans and opinions in times when facial expression and gesture were important reinforcements to their emerging language. Nights were still dark, and later in the human story, those who could calculate, predict and give meaning to the easily seen yet mysterious cycles and phenomena of the night sky became priests and leaders, and some of their durable stone sky-markers still stand. Nowadays, these may be mere tourist attractions, but in the days of their creators they formed foci for settlement and assembly, during humankind's long transition from hunter-

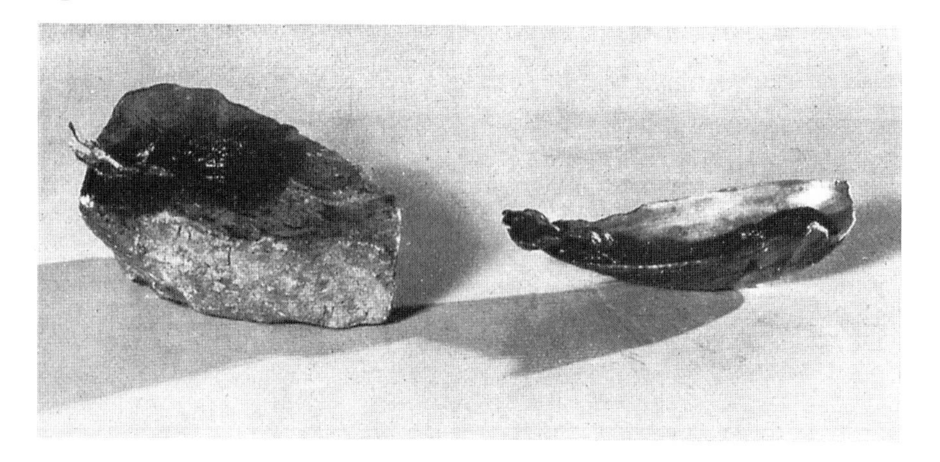

Fig. 1.39. Stone and shell oil lamps (courtesy SSE Museum of Electricity, Christchurch).

gatherer groups to more static and structured communities.

For many thousands of years thereafter, villages slumbered in the darkness of our planet's night, with the occasional candle or burning animal-fat torch to cast a modest glow. Surprisingly little progress was made in the design and construction of lamps between the time of the ancient Sumerians (*c.* 3000 BC), whose hollow-stone, shell and pottery lamps (Fig. 1.39) have been unearthed from the ruins of Ur, and the eighteenth century, when oil or grease burning from a wick still lit human dwellings. Even at the beginning of the nineteenth century, the Eddystone Lighthouse's candle-power was literally just that.

Sleep patterns followed the day–night rhythm more closely, and people slept longer in the past: 9 hours before the invention of the tungsten filament bulb, compared with today's 7½ hours, according to Stanley Coren (*Sleep Thieves*, 1996). The nineteenth century saw the transition from oil lamps to gas lamps and finally to the first electric lighting, with the growth of systematic if comparatively dim street lighting (Fig. 1.40) augmenting the human eye's attempts to fathom the gloom. Sporadic attempts to light streets had been made in earlier centuries: for example, in Paris, in 1367, regulations existed requiring lanterns to be hung in streets, and in 1415 the Mayor of the City of London ordered that all dwellings along the highway should provide an outside lantern every night between All Hallows' Eve (31 October) and Candlemas (2 February). Exceptions were made for the nine nights after the Moon had passed its first quarter, the period when, according to the encyclopaedist Krunitz, "the

Fig. 1.40. A bulky carbon arc streetlamp from the 1880s next to the surprisingly small original filament lamp by Swan (courtesy Eric Jones, SSE Museum of Electricity).

grand light that rules the night can light the street sufficiently, and lanterns are superfluous". In 1588, with the fear of Spanish incursions firmly in the public mind in England as the Armada sailed up the Channel, an order was made requiring city householders to provide outdoor lamps. The penalty for disobeying this edict was death, though, after the passing of the Spanish threat, compliance was much reduced by the substitution of a fine of one shilling.

The first attempts at large-scale gas illumination in several countries were due mostly to the efforts of Murdoch in 1779, but it was well into the nineteenth century before purification of the gas and better burner designs led to Welsbach's incandescent gas lamp, which gave light from the heated mantle rather from the luminosity of the flame itself, but this system could still not be considered a very bright source.

It was an English physicist and chemist, Sir Joseph Wilson Swan, and an American physicist and inventor, Thomas Alva Edison, working independently, who put an end to humanity's natural nights in 1878 by perfecting, almost simultaneously, sealed, bright incandescent-filament lamps, powered by electricity. Swan first demonstrated his lamp (Fig. 1.41) to the Newcastle Chemical Society in December 1878. Within an elongated bulb, he had surrounded the incandescent carbon filament with an almost perfect vacuum, the fruit of his collaboration with C.H. Stearn, who was working on the creation of vacua with the aid of an expert glassblower, Frederick Topham.

In the USA, Edison had also been working on a filament lamp. Experimenting with all manner of filaments, including carbonised cotton, and bamboo,

Fig. 1.41. Bob with a replica of Swan's first filament lamp (courtesy SSE Museum of Electricity).

specimens of which he had explorers send to him from all over the world, Edison produced a lamp in October 1879 which worked for 45 hours. After a period of rivalry, Swan and Edison sensibly pooled their ideas and resources, and the Edison and Swan (later Ediswan) United Electric Light Company was formed in 1881.

When electric lighting is mentioned, most people will have in mind the descendant of the nineteenth-century Ediswan bulb: the filament lamp which hangs from most ceilings in the electrified world. On the streets, however, mercury and sodium discharge lamps, within which gas is excited to glowing by electrons passing through, swept filament types away in the 1930s, thanks to their longer life expectancy and greater efficiency. Denis Crow, former curator of the CU Phosco Museum of Street Lighting, made the interesting point that "with oil, wax, gas and filament lamps, light was a by-product of heat, whereas with discharge lamps, the opposite argument could be made".

The original mercury discharge lamps, dating from the early 1930s, gave between two and three times as much light as equivalent filament lamps; and the sodium vapour discharge lamp, which researchers at Eindhoven in Holland did much to improve, gave four times as much light. The negative characteristic of discharge lamps, though, is that they require auxiliary gear to help start them, control current and improve the power factor.

The gradual, and undoubtedly popular, spread of public outdoor lighting across the developed world has proceeded at an ever increasing rate since the nineteenth century. Great efforts have been made by many clever people to improve the quality and efficiency of the lamps, whatever the light source employed, in

Fig. 1.42. Poorly directed emissions: much of the light misses the church.

Fig. 1.43. Well-directed emissions: the supermarket engineers who commissioned this car park lamp specified downward light only, and screens to prevent light intruding into neighbouring windows.

order that those who wish to venture out at night have an evenly lit environment, and can discern the nature and colour of the objects within it. One aspect of lighting design, however, which seems to have been the "poor relation" for most of that time, is directionality (Figs. 1.42 and 1.43).

The unquestioning approval which the blessings of early street lighting inspired in our forebears has become a habit. Before the late 1980s few people were ready to condemn the excesses of overlit and garish surroundings, assuming perhaps that the discomfort caused by glare was inevitable, normal, the price they had to pay for a lit environment at night (Fig. 1.44). As the stars began to go out, magnitude by magnitude, over great cities and smaller towns; as bedrooms began to be filled with light even with their curtains closed; and as the first of the legions of "Rottweiler" floodlights began to bite in both town and countryside, ousting the traditional more modest and welcoming porch light, little protest was made: the uncomfortable cost, perhaps, of progress (Fig. 1.45)?

Fig. 1.44. Glare dominates the environment in this photo taken on the outskirts of London (photo: Edward Hanna).

Adverse Impacts of Poor Artificial Lighting

Apart from the obvious waste of energy, money and resources which poorly aimed and overbright lighting causes, its adverse impacts may be classified as follows:

- **Cost to the environment** To produce the electricity for misdirected and superfluous light, more fossil fuels are burned in power stations than would otherwise be burned, extra greenhouse gases are produced, and more atmospheric pollution created.
- **Skyglow** This is the visible glow caused by scattering and reflection from atmospheric aerosols and droplets, even on what seem to be the clearest of nights, veiling the stars and degrading the environment above (Fig. 1.46).
- **Light trespass** Light pollution is not just an astronomers' problem. The quality of many people's lives, whether or not they observe the stars, is seriously affected by neighbours' insensitive lighting, or the glare from road lamps, needlessly entering their property (Fig. 1.47). And humans are not the only species troubled by stray light.
- **Glare** Overbright and poorly directed lights can dazzle or discomfit those who need to see, conceal-

Fig. 1.45. Most of the light from this car park floodlight will go into the sky (photo: Mike Tabb).

ing rather than revealing. Much sports floodlighting (Fig. 1.48), and the cheap and vastly overstated 500 W "security" light, so common in domestic use, often fall into this category.

The **cost to the environment** of wasted light, in terms of wasted fuel and greenhouse gas production (notably

Fig. 1.46. Skyglow over a small town.

Fig. 1.47. Reducing light trespass: the luminaires, as can be seen from the pattern of light in the trees, illuminate the road rather than the houses alongside (courtesy D.W. Windsor Ltd).

carbon dioxide) is easily understood and self-evident, and needs little extra comment. The fact that a 100 W bulb left burning though all the hours of darkness in a year causes about a quarter of a ton of carbon dioxide to be emitted by the power station gives pause for thought; and while the current trend is for the use of energy-saving light fixtures within the home, it is certainly worth insisting to any agency encouraging this trend that they must not forget the lights *outside* the house, which can often be much brighter and more wasteful than any indoors.

Skyglow has already been defined. Before about 1950, where street lighting existed, it was a common practice to turn the lights off at about midnight. The relatively small numbers of luminaires in towns and cities meant that, although their designs often allowed emissions to escape above the horizontal, it was usually possible to appreciate the night sky from many places in town, even when lights were on (Fig. 1.49). This may have involved a short walk, but, in those days, when walking was a far more readily undertaken activity, and fear of crime had not begun the climb to its present inflated level, this would have seemed acceptable.

At the beginning of the twenty-first century, the UK street light population has reached approximately seven million (Fig. 1.50); in the USA and Canada, there are

Fig. 1.48. Sports floodlights glare into the eyes of approaching drivers.

more than twenty million roadlights. Most areas of large cities are intensively illuminated, with growing numbers of lights along the roads between those cities. Town-dwellers holidaying in the countryside are often surprised by the fact that some villages are nowadays as brightly lit at night as their main streets at home, and the high-powered security lighting of nervous cottage-holders winks on and off just as much as it does on the urban scene. A principal reason for this is that many of these settlements now have populations composed mostly of ex-"townies", who are happy to sacrifice their new and unaccustomed view of the night sky for the easy reassurance of the urban street furniture they have left behind. Wasted light in the countryside, which Libby Purves described in *Town and Country* (1999) as "this rural plague" (Fig. 1.51), results in the sad fact that, even far away from big towns, many children are now growing up without the opportunity to see the Milky Way, or much detail in the starry sky.

Light trespass, the unwelcome spilling of light beyond the boundaries of the premises within which it is emitted, causing annoyance to people in neighbouring premises (Fig. 1.52), is not yet generally recognised as an actionable nuisance in law. The term originated in the USA, where local lighting ordinances are in effect in some towns and counties.

Fig. 1.49. The city of Bath by night, 1950s and 2000 (photo: Mike Tabb).

Light is not yet considered by most administrations, national or local, as a potential pollutant. Noise, smoke and vicious dogs may well lead to a visit from a public health official empowered to act against these nuisances, but light is not normally actionable* at this level (but see Appendix 4).

---

*However, in at least two recent cases in British courts, successful challenges have been mounted against the originators of stray light, because of its effects on the quality of human life (*Bonwick vs Brighton and Hove Council*) and on the behaviour of wildlife (*Stonehaven and District Angling Association vs Stonehaven Tennis Club*).

Fig. 1.50. One of Britain's seven million streetlamps about to be installed: this one is a full-cut-off design destined to cut down stray light near my observatory (photo: Pam Mizon).

Law lecturer Francis McManus of Napier University discussed the subject at a light pollution seminar organised by the National Society for Clean Air and Environmental Protection (NSCA) in November 1999. He stated that "there is no doctrinal reason why light should not be considered a pollutant, and a nuisance in

Fig. 1.51. Urbanisation of the countryside near John O'Groats (photo: Bill Eaves).

Fig. 1.52. Light trespass: light spill from car park lamps in a rural area illuminates a room in a nearby house (photo: Richard Murrin).

law". In a seminal article, *Light Pollution: a Review of the Law*, in the *Journal of Planning and Environment Law* (January 1998), lawyer Penny Jewkes wrote: "Environmental protection is the sum of small concerns; this is the essence of sustainable development, which requires that decisions throughout society are taken with proper regard to their environmental impact. The planning system goes some way to achieving this, but it was never designed to bear the full responsibility for the control of light pollution."

The main reason for the upsurge, reported by Environmental Health Officers, in complaints involving light trespass during the last decade, is not primarily that existing planning law does not encompass or seek to control small-scale outdoor lighting. The simple fact is that nearly all security lights on retailers' shelves have not been designed with a view to trying to restrict their emissions to the premises to be lit (Fig. 1.53), which would involve the addition of shielding, baffles or louvres, and not least the inclusion in their packaging of instructions on sensitive mounting of these devices.

Sales of such poorly designed and environmentally insensitive security lights rocketed during the 1980s and into the 1990s, to the extent that describing them as "ubiquitous" is not far from the truth. Advertisements for these lights stress their value in deterring the approach of criminals, though the debate

Fig. 1.53. A poorly mounted "Rottweiler" light which illuminates premises across the street.

about the truth of such claims is lively. Certainly, premises possessing these "deterrents" are commonly entered, robbed and vandalised at night, even when the lights are working. Our cities are ever more brightly lit, but this does not seem to result in a reduction in crime, and premises in rural areas are regularly broken into whether or not exterior lights are fitted (Fig. 1.54).

By 1993, the passive infrared (PIR) sensor, activating the light if a heat source enters its field of "view", was becoming a common feature of such lamps. Sensor-triggered lamps were to some extent an improvement on their far more wasteful predecessors, which might be left on all night long. However, in many cases, the manufacturer mounts the sensor centrally below the lamp casing, which prevents the lamp being angled very far downwards even if its owner is minded to prevent spillage of light onto neighbouring premises (Fig. 1.55). Until about twenty years ago, many domestic outdoor lighting units had projecting rain-shields fitted above the lamp housing, because their makers were not obliged to fit glass to the front of the lamp. Now that glass is compulsory as a safety screen, the shield, which would prevent much upward light spill, has been done away with.

Fig. 1.54. Members of the Wessex Astronomical Society lose their once dark observing site in the New Forest to a new 1000 W security light on a nearby cottage. If there were burglars breaking in, how much would we see?

There is good evidence that the quality of life of those affected by unwanted light from neighbouring premises is often greatly diminished. One correspondent (a non-astronomer) who wrote to the British Astronomical Association (BAA) in 1994 described how he had suffered from lack of sleep for a prolonged period because of the domestic floodlights of two neighbours, which were on at most hours of the night and shone brightly through his bedroom window. His request for their modification having been politely but firmly turned down, he was forced to sleep in another room, but was still troubled by the light, the situation being aggravated by the fact that his job required him to get up at a very early hour. In the end, having changed his job as a result of interrupted sleep, he solved the problem at great personal expense by moving house. The BAA's postbag shows that he is not unique in his predicament. In the mid-1990s the BBC consumer programme *Watchdog* examined the subject of light trespass caused by poorly aimed security lamps, and recounted similar stories, including the case of a couple who had even considered emigrating if they could find a country with more stringent nuisance laws! Ill-directed and overbright lighting can cause as much inconvenience and stress as other pollutants, and complainants driven to seek solutions in the courts ought not to be treated as "hypersensitive", or be

Fig. 1.55. An outside lamp with its sensor mounted beneath, making it impossible to angle it down further.

advised by a thoughtless judge – and it was really said – to "fit thicker curtains".

It should be remembered that we tamper with our age-old day/night responses at our peril. Light is the main agent in synchronising circadian rhythms in nature, and human beings, still conditioned by millions of years of evolution to a regular cycle of dark and light, can certainly suffer from an excess of either. Light therapy has been used to good effect in cases of seasonal affective disorder (SAD), a depressive reaction to the dull days of winter; but, as Dr David Avery, professor of psychiatry and behavioural sciences at the University of Washington (WA) School of Medicine, pointed out in a definitive, independent paper on the subject of light as therapy in 1999 (see bibliography), we need a balance between light and darkness. The paper recommends a five-point light exposure regime, stressing the importance of a regular dark-light cycle and the avoidance of sleep deprivation. Dr Avery states that, if these principles are not adhered to, consequences would not be serious for many people, but for vulnerable individuals, such as those affected by SAD, the opposite might be true. Decreasing our exposure to light in the evenings is as important as receiving adequate amounts of light during the day, so closing the curtains at night if security or street lights shine into the bedroom is advisable. Animals threatened, and sometimes even killed, as a result of stray light have no curtains to pull. There are large numbers of reports of its deleterious effects on many different species: insects, birds, fish, reptiles and mammals. The few examples cited here are representative of the problems that countless millions of our fellow-creatures on this planet have because we do not aim lights properly or use correct wattages:

- British Astronomical Association member Robin Scagell has carried out a nationwide survey of Britain's glow-worm population. These intriguing insects (they are not worms) glow like tiny green LEDs in the grass on summer nights, trying to attract a mate. They need only a regular food supply and a dark environment to thrive (Fig. 1.56). Their food (mostly tiny snails) is still available – but Robin believes that the encroachment of artificially bright skies over their habitats is the cause of a rapid decline in glow-worm numbers over the last thirty years.
- Baby turtles in many areas of the world emerge, when tide and Moon are right, from the eggs which their mothers have laid beneath the sand of carefully selected sandy beaches. The hatchlings head for the light of the Moon, which, because of the orientation of the beach selected, will lie out to sea. If the local disco or car park lights along the beach road outshine the Moon, though, the babies will turn the wrong way and scrabble inexorably to their doom inland. A detailed report on the problems

Fig. 1.56. Glow-worms in the foreground grass battle desperately to be seen against the light from a newly-lit main road nearby (photo: Robin Scagell, Galaxy Picture Library).

turtles face with lighting can be found on www.state.fl.us/fwc/psm/turtles/lighting/index.htm.

- In *New Scientist*, March 1995, Fred Pearce considered the plight of exhausted songbirds, singing all night long as security lights constantly triggered their dawn response, the rhythms of millions of years overturned by thoughtless lighting; and the *Audubon Magazine* of April 2000 carried a report from the American Bird Conservancy that at least four million birds die every year in the USA (and the figure may be much higher) because they fly into illuminated towers. There are similar reports of large numbers of sea birds crashing into fishing vessels using powerful floodlights. More information on the effects of artificial lights on birds is on www.abcbirds.org.

Glare is caused by light which is spilled from luminaires in any direction to cause discomfort, distraction or inability to properly see what the light is supposed to be illuminating. Lights which *conceal* rather than *reveal* defeat the whole purpose of lighting, and can truly be called "anti-lights" (Fig. 1.57).

Glare is the most safety-related aspect of light pollution. This effect need not be dazzlingly bright to affect the observer adversely. A simple experiment to assess the impact of glare is to drive along a brightly lit road at night, keeping one's eyes on the road but also being

Fig. 1.57. Glare: a poorly aimed light in rural Northern Scotland (photo: Bill Eaves).

mindful of the effect of the brightness of the road lights; then, lower the car's sun visor, restricting upward vision, so that the light sources (bulbs) cannot be seen directly, and try to gauge any difference in comfort. If you feel more comfortable, the lights are emitting an unnecessary and potentially distracting amount of glare. According to the recognised American authority on standards of illumination, the Illuminating Engineering Society of North America (IESNA, sometimes referred to as the IES), glare may be categorised as follows:

- *Blinding glare:* a glare so intense that, for an appreciable time after the stimulus has been removed, no object can be seen or easily distinguished.
- *Disability glare* (veiling luminance): glare causing reduced visual performance. Drivers in cities are confronted with ever-changing and conflicting light sources, many of them bright enough to cause disability glare: pupils constrict, trying to adapt to the brightest sources, and ability to see into shadowed areas is diminished. Poor and impatient driving may be the result. Disability glare is a typical result of an outward-facing 500 W security lamp being triggered by your presence as you approach the owner's premises – what a welcome!
- *Discomfort glare:* glare producing discomfort or annoyance without necessarily interfering with visual performance. The IESNA confirms that discomfort glare may cause fatigue, with safety implications for drivers, for example.

Modern full cut-off luminaires (FCOs), which have a flat glass sheet, or a slightly convex bowl, beneath a light source housed well up within the casing, emit light only below the horizontal (Figs 1.58a/b/c). This assumes, of course, that they have been correctly installed with the glass parallel to the ground below. FCOs are hardly seen from a distance, and illuminate the road while reducing, if not completely eliminating, glare into the eyes of approaching road users. Many new luminaires have a very shallow curving glass beneath the casing, transmitting light sideways and downwards, to avoid the effect of constant "flashing" into drivers' eyes caused by the sudden staccato appearance of the light sources as the vehicle approaches them.

Floodlights illuminating commercial premises and sporting venues are often a source of glare to passing traffic, but the effect is most commonly experienced

Fig. 1.58a FCO road luminaire with careful optics, designed for residential streets (courtesy D.W. Windsor Ltd).

Fig. 1.58b FCO in profile; this type is increasingly seen on Britain's main roads (courtesy Urbis Lighting Ltd).

Fig. 1.58c FCO with multiple lamps, often used on roundabouts and busy road junctions (courtesy Siemens Ltd).

from incorrectly mounted and overbright domestic security lamps, as described above.

Britain's Institution of Lighting Engineers (ILE) is the highly respected guideline body for lighting professionals in the UK. In 1993, in collaboration with the CfDS, the ILE published the influential *Guidance Notes for the Reduction of Light Pollution*, since twice revised. These recommendations will be referred to several times in this book, and they are reproduced in full as Appendix 3. In the *Guidance Notes*, the ILE states that: "For domestic and small-scale security lighting ... passive infrared detectors can be used to good effect, if correctly aligned and installed. A 150 W (2000 lumen) tungsten halogen lamp is more than adequate. 300/500 W lamps create too much light, more glare and darker shadows. All-night lighting at low brightness is equally acceptable. For a porch light a 9 W (600 lumen) compact fluorescent lamp is more than adequate in most locations." So the experts are on the side of those who appreciate what well-confined, quality lighting means for the environment (Fig. 1.59).

Fig. 1.59. Low-power, shielded security lighting illuminates a porch and garden adequately, without glare or trespass onto neighbouring properties, using two 10 W sources (courtesy IDA).

Security Lamps and Crime

Ignoring the experts' well-publicised message, retailers continue to offer 300–500 W domestic security lamps,

two or more than three times brighter than the *ILE Guidance Notes* recommend, 150 W being "more than adequate". IESNA definitions of glare seem irrelevant to the manufacturers of such devices, which almost never include any instructions on mounting angles and the avoidance of light trespass, skyglow or glare. When bolted to an outside wall, they are often aimed indifferently towards the general area to be illuminated, with their front glass vertical or near-vertical, allowing as much as half the light to be emitted above the horizontal, across the street, well beyond the premises, and also into the sky. If allowed to shine all night, they represent a major environmental problem, and use far more electricity than would a light of a more sensible wattage. The presence of such lights advertises the existence of the premises which they are supposed to protect, especially if they would normally not be seen from a distant road. When they are switched on, or, if fitted with a PIR sensor, are triggered, the resultant glare (Fig. 1.60) has consequences beyond the erasure of the night sky and the increased totals on the electricity bill. Potential witnesses to any criminal activity are dazzled by powerful lamps pointing outwards onto the street, deep shadows in which malefactors may hide are created, and an intruder is supplied with a light source, paradoxically far less likely to cause suspicion in the

Fig. 1.60. Security lights shining into the eyes of approaching drivers.

mind of any right-minded person in the vicinity than a hand-held, moving torch in a darkened area. The IDA reports, in its *Information Sheet No. 54: Dark Campus Programs*, that "school districts across the US are turning conventional wisdom on its head by turning off lights in school grounds ... It seems to work well ... When everyone gets used to dark school grounds, lights of any kind will arouse suspicion". B. A. J. Clark (see bibliography) reports that during the power black-out that affected Auckland (New Zealand) for several weeks in early 1998, criminals almost deserted the darkened streets, a police inspector remarking: "It's been almost a crime-free zone. The normal levels of muggings, violence, fights, burglary and robbery have just not happened."

Martin Morgan Taylor, a law lecturer at de Montfort University, summed up the defects of glaring security lighting in a few sentences in his article "*And God Divided the Light From the Darkness – Has Humanity Mixed Them Up Again?*" in the journal *Environmental Law and Management*, January 1997:

Most domestic burglaries occur during the daytime, when light is not a deterrent. In addition, the use of such lighting is often predicated upon the belief that a passer-by will telephone the police or even intervene should he see the commission of a crime. This is known as "passive surveillance". However, most people will not intervene ... so that no amount of lighting will make a major difference ... Security lighting is frequently placed in secluded rear aspects of properties, to deter criminals. However, if a criminal sets off the light, it means that he is already half committed to the crime, and is unlikely to retreat. More importantly, however, lights in secluded areas are just that: no one can see what the criminal is doing, but he has a "courtesy light' to illuminate his activities (Fig. 1.61).

The U.S. National Institute of Justice report (Sherman *et al.*) on crime prevention, presented to Congress in early 1997, stated in its section on "Conclusions for Open Urban Places":

> We can have very little confidence that improved lighting prevents crime, particularly since we do not know if offenders use lighting to their advantage. In the absence of better theories about when and where lighting can be effective, and rigorous evaluations of plausible lighting interventions, we cannot make any scientific assertions regarding the effectiveness of lighting. In short, the effectiveness of lighting is unknown.

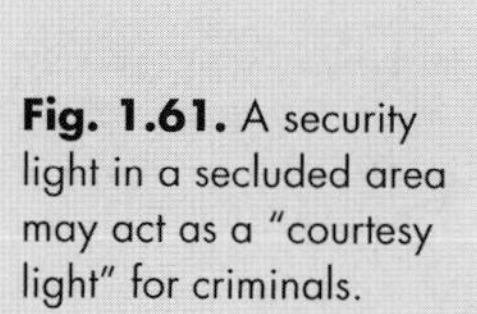

Fig. 1.61. A security light in a secluded area may act as a "courtesy light" for criminals.

In a similar vein, a review by M. Ramsey and R. Newton of studies concerning lighting and crime was published by the Home Office Crime Prevention Unit (HOCPU) in 1991, under the title *The Influence of Street Lighting on Crime and Fear of Crime*. One conclusion was that, although the public believes lighting and crime prevention are linked, there is no real evidence to prove this. The then Deputy Under Secretary of State at the Home Office, I.M. Burns, summed up the general tenor of the report by stating that fear of crime might well be allayed by lighting, but the incidence of crime was not so easily linked to improved lighting. B. A.J. Clark, senior research fellow in the Department of Optometry and Vision Sciences at the University of Melbourne, Australia, and a prominent amateur astronomer, wrote in his report (1999–2000) for the guidance of Melbourne City Council: "Crime is a social problem, not a lighting problem." He makes the interesting point that "intense or continuous lighting is generally unnecessary for personal or property security, and may even encourage crime because fear of crime is allayed and commission is facilitated".

Many studies (not always independent of vested interests) have been made on both sides of the Atlantic on this subject, some showing that increased lighting deters (or displaces) criminals, and others the opposite effect. *If* light has little effect on criminal motivation, then those who play on the public's fear of crime to sell them glaring security lamps are not only harming the environment, but are also alarming people to an unnecessary and possibly illegal degree. If some criminals are deterred by lights, then their quality rather than their actual presence will be the determining factor. The *Home Office Crime Survey*, published in October 2000 and based on the experiences of victims of crime, suggests that premises which have security lighting are as likely to be broken into as those without it. Might it be inferred from this that the presence of the lights has in some cases attracted the crime?

In 1995, a new, brightly-lit Super K-Mart parking area in New Haven became one of the highest single crime foci in the state of Connecticut. The *New Haven Register* reported fifty-four incidents in December, stating that the K-Mart parking lot had been dogged by criminals since it was opened, and that this was not unusual for such large retail parking areas. David Crawford of the IDA echoes this story: "Another recent quote was about lighting at a car storage area, unoccupied at night, but not much lit up. It was near a major highway, and burglars would pull off, cut a hole in the fence, grab parts and leave, fast. The police finally caught one, and asked 'Would better lighting help?' The burglar replied: 'Sure, I could get in and out a lot faster and not get caught.'"

An interesting case, illustrating unreasoning faith in light as an infallible crime deterrent, was reported in the *Liverpool Echo* in November 2000. Night-time vandals smashed fifty-five of the ground-level floodlights, which were switched on at the time, placed around the city centre's vast temple-like St George's Hall. In the same report, a local councillor was quoted as saying that she believed that lighting in Liverpool's city centre was a vital measure to keep crime out of the city (!).

It is difficult to decide, in the face of conflicting and not always independent evidence, whether security lighting actually does, in general, what it is supposed to do, but the word "security" very often seems ill chosen. It is perhaps more productive to examine the motivation and needs of the offender rather than the claims of retailers.

The HOCPU review mentions an independent study by Bennett and Wright (1984) entitled *Burglars on Burglary*, in the course of which three hundred experienced burglars were interviewed, and asked about their decision-making processes. According to the survey, the excitement of risk-taking was perhaps their greatest motivation to break into premises, and they routinely took into account the presence of neighbours or dogs, the extent to which the property was overlooked, and the possibility of intervention by passers-by, though as many as half stated that they were not necessarily deterred even by these factors. Signs of occupancy were the only truly powerful deterrent, according to nine out of ten of those interviewed.

Hardly any mention was made, though, of lighting conditions. HOCPU also carried out a survey of street robbers, concluding that neither darkness nor lighting were significant factors in the minds of offenders when deciding the time and place of their offences. Manchester Probation Service (1990) carried out research involving a hundred thieves whose targets were cars or their contents. Only one cited unlit parking places as an incentive to commit the crime. There are arguably crimes and incivilities which need a good level of light for their commission: graffiti artists and vandals, whose satisfaction comes from observing the results of their activities, do not work in darkness, for example.

The lighting and crime debate has been an aspect of the wider light pollution debate for many years, but it is, in a sense, a "red herring". In a democracy, those who want to light their premises cannot be denied the right to do so, and astronomers, *who have the same lighting needs as every other citizen*, would seem churlish if they advocated doing away with street lighting or domestic exterior lighting, whose benefits they too may enjoy. The *amount and quality* of light produced at night, the *need* for it, and *where it goes*, are the important things here. Those elected to protect the environment, not only by day but also by night, should be taking steps to ensure responsible lighting practice, both public and private, by legislation and education, with a view to preventing those who have a perfect right to light their premises from harming the environment and disturbing the legitimate pursuits and lifestyles of neighbours.

Lights that are too bright for the task certainly harm the environment, but nobody has yet calculated the ever spiralling monetary cost of these installations. Added to a 1993 estimate (CfDS) of £53 million wasted skywards annually by Britain's streetlamps alone, or to the estimated $13 million that New York's streets throw into the night sky every year (D. Taleghani/IAU), the cost is staggering.

In summary, the adverse effects of poor quality lighting are plain for all to see, and a source of annoyance and concern to many. Those who wish to achieve the worthwhile goal of convincing legislators, local administrators, manufacturers and installers to follow good lighting practice should marshal their arguments carefully. Before presenting the case for improvement in this field to those who can really change things, it is useful to have "facts and figures" ready, and to adopt a scientific as well as a (rightly) emotional approach in favour of the environment and against wasted energy. The next section attempts to provide such facts and figures.

Quantifying the Problem

The measurement of the amount of skyglow above a given location is not easy, depending as it does on many variables. It is not just the number, brightness and distance of the light sources on the ground, and the percentage of their emissions which escapes above the horizontal which must be taken into account, but also upward reflectivity of surfaces in the vicinity of the sources, and the state of the atmosphere. Reflectivity is defined as a measure of the ability of a surface to reflect radiation, equal to the reflectance of a layer of material sufficiently thick for the reflectance not to depend upon the thickness. Reflectance (ρ) is simply the ratio of the reflected flux to the incident flux.

Anyone standing on a piece of high ground overlooking a town or city can see that most of the waste upward light is coming from the lamps themselves (Fig. 1.62) rather than from the ground below the lamps – yet a certain amount will always be reflected by any surface (roads, pavements, walls, windows) likely to be found in a built-up, lit area. The condition of the surface will also be a factor in its reflectivity. Is it clean or dirty? Dark or light? Rough or smooth? Reflective, absorbent or semitransparent? Wet or dry? Estimates for this reflected amount average a few per cent of the incident light emitted from the source, but it is impos-

Fig. 1.62. This view of Malvern and Worcester taken from nearby hills shows plainly that most of the waste light comes from the luminaires, not from the ground (photo: Chris Baddiley).

sible to give a standard figure because of all the variables involved. David Coatham, Technical Services Manager at the ILE, said in a radio interview about light pollution in November 2000: "We can control light that goes directly upwards, but a lot of the light that goes down onto surfaces is reflected into the night sky and is very difficult to control." The answer to this, of course, is to put the *right amount of light* for the lighting task onto those surfaces, and the amount (inevitably) reflected into the sky will be minimised. The organisations working to reduce light pollution know that, because of this reflection factor, the total eradication of skyglow is not an achievable goal. Neither can we expect everybody to happily switch off all their lights. The necessary support from an educated public will never come if people think that astronomers and environmentalists want to plunge them all into utter, medieval darkness as soon as the Sun goes down.

Furthermore, the amount of skyglow seen from any location can vary noticeably with atmospheric conditions. Low cloud cover, even if very thin, is quite reflective, and will reveal the presence of sources of upward light beneath it (Fig. 1.63). A Dutch astronomer, H. E. Mostert, even invented a "clear sky detector", a device which detects the waste light

Fig. 1.63. Reflection from a thin veil of low cloud over Edinburgh (photo: Chris Baddiley).

reflected from clouds: when the sky clears, causing the glow to diminish, the device can be programmed to emit an audible signal: possibly the only positive use that has ever been found for skyglow (see bibliography).

At the other end of the spectrum, there can be nights when even normally severe light pollution is considerably alleviated. It is often noticed by regular observers of the night sky that, after rain, the sky may seem clearer than usual, as it contains less suspended dust and other impurities. Very occasionally, when air quality changes due to the constant motion of air masses, sky transparency can increase markedly, with upward light being reflected and scattered to a much lesser degree (Fig. 1.64). But there is no such thing as a completely transparent atmosphere, which would allow light passing through it to continue its journey without encountering particles and droplets.

The IDA has published a formula elaborated by Merle Walker in 1977, with a view to estimating typical skyglow. Walker measured sky brightness for cities in the south-western United States, and arrived at the formula, sometimes quoted as "Walker's Law":

$$\log p = -4.7 - 2.5 \log r + \log n$$

where p is the ratio of sky surface brightness at an elevation of 45°, measured in the direction of the urban

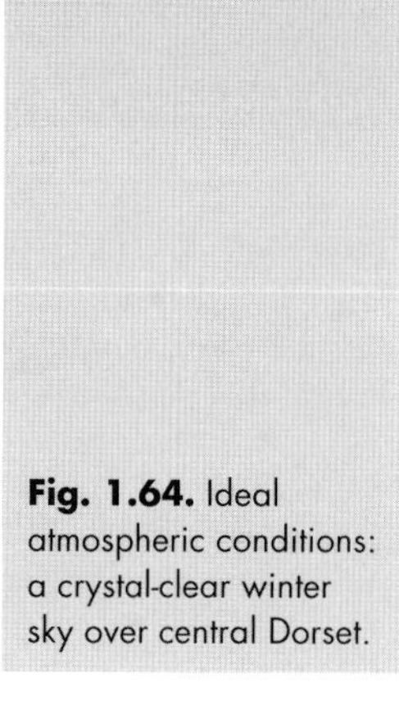

Fig. 1.64. Ideal atmospheric conditions: a crystal-clear winter sky over central Dorset.

source, to the natural background radiation (described in Section 1.1); r is the distance from the source to the measuring site in kilometres; and n is the *total* luminous flux of the outdoor illumination in lumens. The lumen (lm) is the SI unit of luminous flux emitted in a solid angle of 1 steradian by a light source of intensity 1 candela (cd).

For everyday discussion, a simplified formula based on Walker's work has been proposed (IDA):

$$I = 0.01\ Pd^{-2.5}$$

where I is the increase in skyglow level above the natural background at altitude 45° (in the direction of the urban source), P is the population of the urban

Table 1.2. Distances where 10% skyglow is experienced, as a function of populations (source: IDA)

Distance (d):	10 km	25 km	50 km	100 km	200 km
Population:	3160	31 250	177 000	1000 000	5660 000

area, and d its distance in kilometres from the observer. An example given by IDA is:

Example: $I = 0.02$ means that the sky background is 2% above the natural background midway between the horizon and the zenith in the direction of the city, and 1.00 means that the skyglow is double the natural background, a 100% increase. The equation seems to best fit communities where the average value for lumens per person is between 500 and 1000. Large cities emit more light per person, and the amount of skyglow may be larger than the formula shows, perhaps. For example, a city with a population of 1 000 000 that is 100 km from an observing site may produce more skyglow than the 10% the formula indicates.

We can calculate the urban population from which the skyglow will be 10% above the natural background, for a given distance of the observatory from the city (Table 1.2). At this skyglow level, significant sky degradation is beginning.

So, a fairly small town or large, lit village with a population of only 3000 people will cause significant degradation of the sky in its direction from nearly 10 km away. The effect of increasing distance on perceptible skyglow is quoted by IDA, as in Table 1.3:

Remember, the values are for "typical" skyglow. The astronomers' and environmentalists' task is to redefine "typical". How?

1.4 New Lamps for Old

In his *Briefing Sheet on Road Lighting and Highway Power Supplies* (1992), lighting engineer J. Knowles

Table 1.3. Effect of increasing distance on given skyglow level (source: IDA)

Distance (r)	10	20	30	40	50	60	80	100 km
Light level	316	56	20	10	6	4	2	1

Fig. 1.65. The M1 motorway: full-cut-off lamps direct their light down, in contrast to older, glaring lights in the distance (courtesy Urbis Lighting Ltd.).

stressed that the two most important points which need to be taken into consideration to ensure good quality lighting without pollution are well designed luminaires, with modern optical control directing light downwards, and the minimum amount of light for the task. The ILE expressed this more simply in its *Guidance Notes* (1994): "Light pollution can be substantially reduced without detriment to the lighting task" (Fig. 1.65).

Basically, those making, choosing and installing luminaires need to use the capability of internal reflecting surfaces (the optical system) and the lamp glass to direct the emissions to the area to be lit, the mounting height also being relevant here; and the light source within needs to be of the minimum brightness necessary to perform the lighting task adequately.

It is not difficult to find, in the artificially lit environment, luminaires which conform to these ideals, and others which manifestly do not. What are the most common lamp types?

Types of Lamps

The majority of lamps used nowadays in public lighting are of the high-intensity discharge type, which have compact arc tubes. Lamps with longer tubes will probably be of the low-pressure sodium type (see below). Private lighting, which normally involves the floodlighting of premises for commercial or security pur-

poses, often involves tungsten-halogen or metal halide lamps as well as sodium.

(1) Sodium lamps

Most road lamps are of the high-pressure sodium (HPS, sometimes called SON) or low-pressure sodium (LPS, sometimes called SOX) variety. The terms high-pressure and low-pressure are relative, because the gases in the discharge tubes which ionise to form a discharge are always below atmospheric pressure. With a little familiarisation, these two types are easy for the non-technical observer to tell apart. HPS luminaires tend to have smaller casings and light sources, and, being an arc source, an HPS lamp can cause considerable glare if not well shielded. LPS lamp tubes tend to be long (sometimes more than a metre: Fig. 1.66). HPS light has been variously described as pink, pinkish-orange, or golden white, but its overall effect is of a paler cast than LPS emissions, which are of a bold yellow-orange hue. Another indicator is that monochromatic LPS light gives a rather "muddy" and uniform look to surfaces, with poor rendition of their actual colour, while HPS light gives a truer perception. HPS lamps typically use higher wattages than LPS, but may be much more efficient when combined with a modern high-performance luminaire.

The length of LPS lamps makes them difficult to enclose (though there are small numbers of flat-glass, well-enclosed but expensive LPS types on the market),

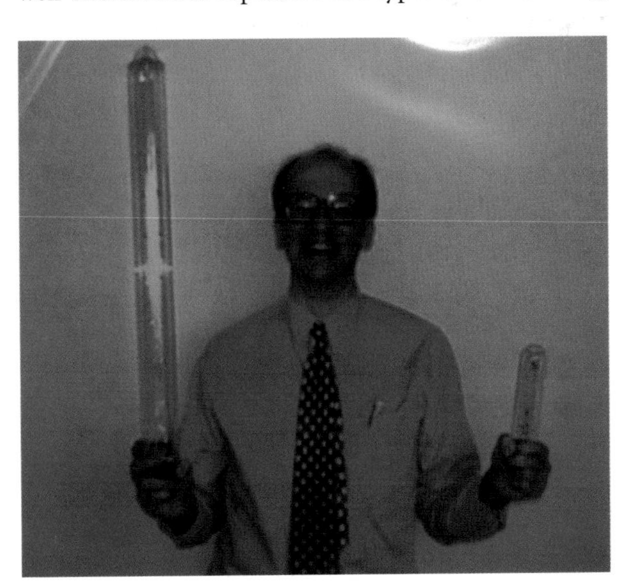

Fig. 1.66. The enormous light source of a 135 W low-pressure sodium lamp with its much smaller 150 W high-pressure sodium counterpart (photo: Patrick Baldrey).

Table 1.4. Economics of light source selection (source: J. Knowles/Hants CC Energy). The table assumes that lights will be on from dusk till dawn, controlled by a photocell. At the time of writing (1/2001), £ = $1.40

Nominal lamp wattage (W)		Typical output (lm)	Annual electricity costs: £	Other maintenance costs: £	Annual operational costs: £
LPS	35	4300	6.72	15.58	22.30
	55	7150	8.91	16.32	25.23
	90	12250	14.75	17.49	32.24
	135	21200	24.01	19.24	43.25
HPS	70	5850	11.83	14.00	25.83
	100	9500	16.65	13.48	30.13
	150	14000	24.68	14.00	38.68
	250	25000	40.31	14.92	55.23
	400	46500	63.68	16.00	79.68

and this means that their light is not as easy to direct as light from a smaller HPS source. Their monochromatic emission is preferred by some astronomers, who rightly state that it is easier to deal with by using a sodium filter than is HPS light, with its many combined wavelengths. This is a complex debate. Local councils and especially the police usually advocate HPS because of its better colour rendition, though HPS lamps may be more expensive to run (see Table 1.4); and the enormous majority of those who enjoy the sight of the night sky, astronomers or otherwise, will never possess such filters. For them, the direction in which the light is made to travel, and its intensity, are the most important factors, and HPS scores over LPS in these aspects.

In both the USA and the UK, the trend since the mid-1990s has been for the replacement of LPS with HPS or "whiter-light" lamps, and for HPS lamps, many with better light control (Fig. 1.67) to be preferred in new schemes. Astronomers with observatories who really desire LPS lamps in their locality might be encouraged by the pronouncements of the Vienna-based Commission Internationale de l'Eclairage (CIE), the international guideline body on lighting practice, who have published guidelines about lighting near astronomical observatories. In the early 1990s the CIE proposed that exceptions might be made in HPS lighting schemes, with LPS types used near observatories. More information about the CIE and its publications may be found in the bibliography.

Fig. 1.67. The cone of downward light from a cut-off HPS lamp (photo: John Chapman-Smith).

Light sources are selected usually on the grounds of economics, involving costs not only of luminaires, but also of installation, maintenance and energy consumption. Information from US sources on running costs of various lamps can be found on the IDA website, information sheet 4 (see Bibliography). As an example for sodium lights alone, we can cite Knowles, who has produced Table 1.4 showing costs of sodium lights, updated by the author for the year 2001, using information from Hampshire County Council energy manager Alan Dowdell.

(2) Mercury lamps

In limited applications, for example where very accurate colour rendering is needed, high-pressure mercury vapour (MBFU) lamps are used. Usually seen as a "white light" source, such lamps are often used for illuminating traffic signs, though they produce typically only 50 lumens per watt of blue-white light, half that of LPS lamps. The IDA reports that some Arizona lighting ordinances (e.g. in Tucson and Pima County) prohibit the installation of new mercury lamps, on the grounds of their inefficiency. Their relatively low efficacy in converting electrical power into light is to some extent compensated for by their longevity.

Low-pressure mercury (MCF) types are occasionally met with for some tasks where a lower level of light is required, but mercury lamps generally have fallen out of favour nowadays in most outdoor applications.

(3) Metal halide lamps

A bluish-white high-pressure discharge source, and more energy efficient than mercury lamps, metal halide lamps are probably the primary sources used nowadays for white light. As the use of white light for functional and decorative purposes grows, they are increasing in popularity at the expense of sodium lamps. They are common in such applications as car park lighting and the illumination of commercial premises, though wattages used are often too high, giving a harsh, overlit effect which the IDA has dubbed "prison-yard" lighting. If not well shielded, they can cause considerable glare and deep shadows. Metal halide types have shorter lives than high-pressure sodium lamps, and are at present more expensive.

(4) Tungsten-halogen lamps
These white-light arc or filament lamps, only a little more efficient than the incandescent bulbs which light most of our houses, are the main culprits when glare is experienced from the outside walls of domestic premises. Relatively short-lived, they are the preferred light source for floodlights and spotlights, and typical values for domestic security floodlights are 300 W and 500 W, both far too bright for the task of illuminating the average small garden or driveway. Drivers and passers-by dazzled by a domestic exterior light a hundred metres away are usually victims of a tungsten-halogen lamp.

(5) Fluorescent lamps
We are all familiar with the long low-pressure discharge tubes of fluorescent lighting, more commonly associated with interiors. The dimensions of fluorescent lamps mean that the direction of their emissions is not easily controlled, though glare is less of a problem. The development of compact fluorescent sources means that these lamps may be seen more often in the future for outdoor applications, though their use will probably be restricted to areas such as residential streets in rural areas where high lighting levels are not required; a typical wattage for a compact fluorescent source is 55 watts.

(6) LEDs
LEDs are creating a lot of interest in lighting circles nowadays, and there is considerable investment in their development. It is estimated that the lumen output of LEDs is doubling every two years, and in the near future they may well be in use for outdoor applications. One problem, however, is the difficulty of white light production from LEDs.

Globe Lights (Spheres)

Leave through the back door of 19 New King Street, Bath, and you step into the garden where William Herschel doubled the size of the solar system overnight, on 13 March 1781, when he discovered the planet Uranus through his "7-foot reflector" (the figure refers to the focal length). The great man's house, where he lived with his sister Caroline, also an accomplished astronomer, is nowadays a splendid museum. It has recently been beautifully refurbished, but the garden was long ago shortened by the intrusion of later building. Opposite the garden door stands a statue of William and Caroline, he staring sternly into the heavens, towards the place where the new planet first caught his eye at the telescope, while his sister carefully records observations (Fig. 1.68). Were the Herschels to return to New King Street today, though, there would be little chance of them discovering a new planet; nor would they see many stars over the far end of their garden, for there is now a supermarket, a shopping

Fig. 1.68. The statue of William and Caroline Herschel in their garden, close to the spot from which Uranus was discovered (photo: Mike Tabb).

Fig. 1.69. Globe lights now illuminate the area behind the Herschels' garden.

complex and a car park there; and these are at present lit by globe lights (Fig. 1.69), some uncapped.

Seldom seen along roads, globe lights, sometimes called spheres or opal spheres, were becoming very popular in the 1980s and early 1990s for public areas such as car parks, walkways and shopping precincts.

Fig. 1.70. A globe light painted black on one side in an attempt to retrieve darkness for an upper-storey bedroom.

Globes contain light sources of various kinds, often with little or no shielding, and more than half of the emissions from these (unshielded) sources will go skywards, since the sphere is mounted on a cradle atop the supporting column, which prevents light from passing directly downwards from the lamp to the area around the base of the column. A person standing close to such a lamp is, paradoxically, not as well illuminated as somebody further away. When they are erected close to buildings, much of the poorly controlled light from globes can enter upper-storey windows, and even more of it will, of course, end up in the night sky. Figure 1.70 shows the result of the exasperation people can feel when such uncontrolled light pours into their bedroom window! Globe lights are good examples of luminaires chosen more for their daytime appearance than for their night-time efficiency. The effect of massed globes, on sites where environmental considerations have been sacrificed for decoration, is shown in Figure 1.71. This photo shows skyglow caused by nearly 90 unshielded car park globe lights erected in the fairly small town of Newport, Shropshire, in 1993, in the car park of a large new supermarket. The lights all but obliterated the night sky over the town. In this case, swift intervention by members of the CfDS led to the capping of the lights, at comparatively little expense to the national supermarket chain involved, though it would of course have been more economical to have chosen better lighting in the first place.

Fig. 1.71. Newport, Shropshire: the stars I learned in boyhood are veiled by a supermarket's car-park globe lights.

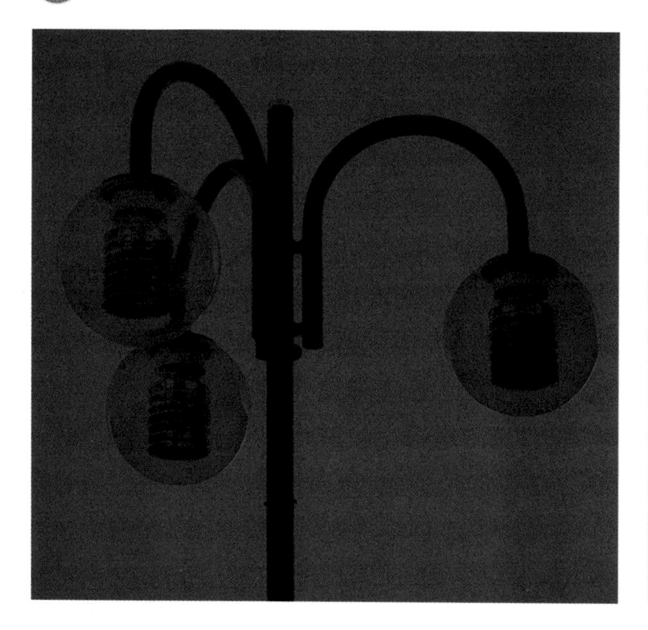

Fig. 1.72. These globes have internal stacked louvres which severely restrict upward light.

Even globe lights can be made to perform in a more environmentally responsible way, if a little thought is given to their design. Figure 1.72 shows internal louvres stacked within the sphere around the lamp, directing all the light downwards, though a small per-

Fig. 1.73. A woman stands near a globe light, and is easily seen (courtesy IDA).

centage will be reflected upwards and outwards by the inner surface of the sphere (see next section). Figures 1.73 and 1.74 show the poor performance of a globe light: as the woman standing near the light moves in towards it, she becomes less visible. Figure 1.75 is a striking example of the poor light distribution from a globe light in a car park: the pool of relative darkness beneath the light is very obvious. Much of the light which should be illuminating the surface below is in fact being thrown upwards into the sky.

Shielding and Directional Adaptation in Luminaires

I don't intend in this non-technical book to go into too much fine detail of how lamps are designed and engineered to shine preferentially downwards – it certainly isn't as easy as it sounds! Most astronomers who are interested in minimising light pollution care more about where the light goes rather than about how it gets there. Anyone who wants to explore the minutiae of luminaire design should contact the IESNA (www.iesna.org), or ILE (www.ile.co.uk) or order the lighting manufacturers' descriptive sales catalogues,

Fig. 1.74. The woman seen in Fig. 1.73 has moved into the less illuminated space beneath the globe light (courtesy IDA).

Fig. 1.75. The fact that most of the light from this car park globe goes up instead of down is excellently illustrated by this IDA photo.

which often carry diagrams of the luminaires' reflective optical surfaces. All that is needed here, in order to give pertinent information which might be useful in discussion, are a few facts about the nature of these reflective surfaces above and around the light source; they are responsible for the intensity of the distribution of the light emitted, while the glass sheet or bowl/globe through which the light passes out into the environment creates further effects.

Optical directors in luminaires can be compared, in a way, to the optics of telescopes. Consider a telescope forming an image of a star. Light from a star enters the telescope as a parallel beam of light, and is imaged to a point. Stars in different parts of the field of view form separately located images at the focal plane of the telescope. An optical system in a luminaire is rather like a telescope working backwards. For a well-directed luminaire, a point source at the focus of the optical system will be imaged to a parallel beam or a deliberately diverging one; the smaller and more point-like the light source is, then the better the directionality of the beam. Luminaire optical systems should be designed to spread the light as evenly as possible over a defined area. The reflector is over the light source and the light source needs to be as small as possible to give the tightest distribution of beam, as in the case of the smaller high-pressure sodium bulbs. This is why such optical systems are rarely used on long low-pressure sodium tubes, though there are exceptions, and low-pressure sodium full-cut-off lights do exist, as mentioned above.

The luminaire's reflector behaves rather like a parabolic mirror in a telescope, turning the upward light from the lamp light into a downward cone. The light that shines directly down from the bulb is similarly restricted by extending the mirrored sides down to below the level of the bulb and to a cut-off point that restricts the sideways light. A further refinement is a reflective spot directly below the bulb, to turn direct downward light back towards the reflector above. This redirects the otherwise most intense part of the beam, to add to the evenness of the illumination.

Asymmetric reflectors are used to direct the beam out towards the centre of the road, with less light falling nearer to the lamp column. Varying degrees of asymmetry are used, according to whether the luminaire has to illuminate the whole width of the road, or just one side.

Cylindrical or globe lamps and bollard lights, often seen in car parks and along walkways, sometimes use louvres to turn the light preferentially downwards. One problem here is that the louvres are sometimes quite reflective, resulting in multiple reflections between adjacent louvres. The beams then spread out from the surface of last reflection. The efficiency of such louvred lamps varies, but they are certainly an improvement on types with no director at all. All glass surfaces introduce an internal reflection of about 10% of the light, which then exits in some other direction. Of course, capping globes with light-proof hemispheres will prevent much upward light, and many manufacturers produce accessories allowing lamps to be "retro-fitted", i.e. altered in design after installation.

Some capped cylindrical and globe lights also use arrays of small prisms to refract otherwise upward components downwards. Again there is the multiple reflection problem, and the scatter from the prismatic screens, if optically very imperfect, is significant, still giving sideways and upward waste light. On more typical road lamps, the design of bowls slung beneath the light source is important: the more curved the glass bowl, the more widespread the beam will be. Some "full-cut-off" lamps have deep bowls, and these produce a secondary image of the light source which is well below the level of the cut-off and can be seen at a range of angles.

Mounting angles are also important in cutting down skyglow. Some shallow-bowl or flat-glass luminaires, which would normally emit little or no light above the horizontal, can cause skyglow and glare if they have been mounted with the glass at an angle well above the horizontal. If your local lighting provider tells you that "non-polluting" lights are to be installed, check that they will be installed at an appropriate angle. If cut-off lights are installed on a slope, will their glass still be truly horizontal?

There is always a "best light" for the lighting task, and any manufacturer worthy of the name ought to be able to give expert advice on which is the best luminaire to use to prevent light spill and skyglow.

Chapter 2
Piercing the Veil: Techniques and Targets

Educating the general public about responsible lighting is a massive undertaking for those who value the environment above. Telling your own neighbours about it can also be a daunting prospect. When the International Dark-Sky Association and the BAA Campaign for Dark Skies began their work in 1988 and 1989 respectively, the concept of "light pollution" certainly existed in the minds of astronomers, environmentalists and the lighting community, but almost no-one else had heard of the term. It is now in the general vocabulary, but the perception of the rôle of light at night is still dominated in many minds by the idea of "the more light, the better". It is quite possible to present a whole range of arguments about economy, effect on crime, and protecting the environment to the owner of a "Rottweiler" light, and still be told that "I feel better with it on, though". It can be difficult to convince nervous "light junkies" of the ill effects they can inflict upon others and upon their wallets, and in Chapter 3, we shall look at strategies for educating about light pollution.

Depending on an astronomer's location, and on the perceptions and degree of co-operation of neighbours, he or she may have a night sky full of interesting and attainable objects, or have to contend with skies flooded with artificial light (Fig. 2.1). Most observers probably experience something in between. Those who have ceased going out at night altogether, defeated by local lights, ought perhaps to reconsider: they should not only be looking for new ways, through campaigning and education, to try to achieve longer-term

Fig. 2.1. The village of Corfe Castle, an island of wasted light in an otherwise dark landscape.

solutions, but also be taking a fresh look at what may still be seen through the veil of wasted light.

2.1 Techniques

The most obvious courses of action for the frustrated observer whose instruments fail to show fainter objects in a tainted night sky are:

- try to filter out the offending light;
- invest in imaging equipment (a CCD camera) which, with the aid of computer technology, will help to pierce the veil;

- observe only relatively bright targets: there are plenty of books on observing the Sun, Moon and planets, but many more remote objects can still be seen through moderate light pollution, with adequate equipment and in calm conditions, using more traditional methods only.

Let's discuss filters and CCD equipment first, and then offer some targets which do not need the high-tech approach.

Filters

There is a bewildering array of advertisements for filters in the astronomical press. Many of them promise that the products will reduce light pollution. Filters, which work by blocking unwanted light and admitting light in a fairly narrow region (bandwidth), can be effective devices, but if the object observed emits a proportion of its light outside the bandwidth, it will be dimmed. Remember always that filters *subtract* light; they do not add it, so they will not "brighten" the objects you want to look at. Figure 2.2 shows some popular LPR filters.

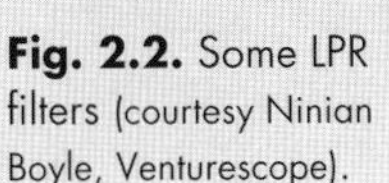

Fig. 2.2. Some LPR filters (courtesy Ninian Boyle, Venturescope).

Light Pollution Reduction (LPR) filters are just that: they are not light pollution elimination filters. They will not magically clear the sky, or restore it to its pristine state, though they can improve your view of certain objects, in dark skies as well as in light-polluted ones. They may not remove all the effects of light pollution, but by reducing them they can improve the visibility of some deep-sky targets, increasing the contrast between them and the background sky. The three types of LPR filters are: broad-band, narrow-band and line (extremely narrow-band) filters.

Broad-band LPR filters are designed to prevent the passage of light from the most common lamp types, namely HPS, LPS and mercury vapour; any light of the same nature coming from an astronomical object will also be removed. Broad-bands have bandwidths typically greater than 30 nanometres, measured at the "half-maximum" or 50% transmission level. Bandwidths for some popular broad-band filters are: Orion Skyglow, 85 nm; Lumicon DeepSky, 68 nm; Celestron A, 47 nm.

Sodium lamps produce yellow light centred on the so-called Na-D line at 589 nanometres, so the reddish or bluish light of nebular objects is not blocked. HPS sources emit a much wider range of wavelengths than LPS. Mercury lamps produce a variety of emissions, mostly below 450 nm and above 550 nm. Deep-sky objects like galaxies and clusters, whose light consists mostly of starlight, are usually little enhanced by LPR filters. The objects which show up better tend to be emission nebulae such as the planetary nebulae (e.g. M57, the Ring Nebula) and supernova remnants (e.g. NGC 6992-5/6960, the Veil Nebula). Some old LPR filters block out only LPS light.

Narrow-band filters, marketed as Ultra High Contrast (UHC) or Ultrablock filters, have bandwidths typically smaller than 30 nanometres. Bandwidth values for some popular narrow-band filters are: Lumicon UHC, 27 nm; Orion Ultrablock, 24 nm.

Line filters are normally of the Oxygen-III (O-III) or Hydrogen-Beta (Hβ) type, and perform well on non-stellar objects. Their bandwidths are very restricted, and they reject nearly everything but one or two given emission lines. Bandwidth values for two popular types are: Lumicon O-III, 11 nm; Lumicon Hydrogen-Beta, 8 nm.

So which filter to choose for your favourite planetaries, star clusters, and galaxies? Table 2.1 (based on

Table 2.1. Filters and performance

Object	Broad-band filter	Narrow-band filter	Line filter
Planetary nebulae	3–4	5–6	6*
Other emission nebulae	3–4	5–6	6*
Reflection Nebulae	2–3	2	1
Stars, clusters	2	1–2	1
Galaxies	2–3	1–2	1

*Normally: some emission nebulae have emission profiles which favour certain line filters. Planetaries tend to be rich in O-III emissions, while other types of emission nebulae are relatively stronger in Hydrogen-Beta.

David Nash's Internet LPR Filter Guide) will help. In the table, performance ratings range from 1 to 6, meaning:

1 Very poor: object looks much worse than it would without the filter;
2 Poor: object noticeably worse than without filter;
3 Fair: object about the same, or even a little better;
4 Good: object normally looks appreciably better;
5 Excellent: object normally looks dramatically better;
6 Outstanding: object normally looks dramatically better even in conditions of low light pollution (e.g. dark rural site).

So, for enhancement of emission nebulae and better subtraction of light pollution generally, narrow-band are the filters to go for. Galaxies, stars and star clusters show better through a broad-band filter, though the benefits are more subtle. Interestingly, airglow, which emits predominantly at 465, 558, 630 and 636 nanometres, can also be subtracted by various filters, which explains why they can also enhance objects, especially those of low surface brightness such as the Veil Nebula in Cygnus, from dark observing sites. Most narrow-band filters will deal with the 465 nm line; all filters block the 558 nm line; filters which perform well on the red 630 and 636 nm lines include the Orion Ultrablock, and to a lesser extent, the Lumicon UHC and the Lumicon DeepSky.

Words of warning: LPR filters are very reflective, so take care to block out stray light from behind and beside you. Eyecups from binoculars can be useful here, or, if you don't mind looking like a nineteenth-century photographer, drape a thick black cloth over your head and the eyepiece. Also, if instead of attaching the filter to your eyepiece, you are holding it between finger and thumb for "blinking" purposes, moving it rapidly to and from the eyepiece to compare views of nebulae, be sure to keep it parallel to the eyepiece lens. Tilting the filter will affect performance, as you can demonstrate to yourself by looking at a light though it and watching its colour change with the angle of the filter as you slowly rotate it.

As in many aspects of astronomy, experience and practice count for much when using filters. Before trying your newly bought filter on a faint, challenging object, try something more attainable. The Great Orion Nebula or the Lagoon Nebula are ideal for finding your way with filters. Let confidence and success come in easy stages.

As an infrequent user of filters (they have a way of escaping from my small observatory (Fig. 2.3), never to be seen again), I am indebted to David Nash (USA) and

Fig. 2.3. My small run-off-roof observatory in Colehill, with its 21-cm/8.5-inch reflector.

Steve Tonkin (UK) for much of the up-to-date advice and information above. Their informative websites are listed in the bibliography. You can find articles on popular filters in *Astronomy*, February 1991, in *Sky & Telescope*, July 1995, on David's website, and in Steve's book *Astro FAQs* (also in the bibliography).

CCD Astronomy

Those able to invest in a CCD (charge-coupled device) camera, used in conjunction with a telescope, automatic guiding equipment and computer back-up, can achieve stunning images of night-sky objects even from light-polluted urban sites. The possibilities opened up by the image processing which CCDs allow have rekindled interest in astronomy for some urban observers. A detailed discussion of how CCDs work is outside the scope of this book; there are excellent books available on CCD techniques (see bibliography).

Computer-processed CCD images taken nowadays from urban areas (Figs. 2.4 and 2.5) can rival photographs obtained in earlier decades at the world's great observatories, and CCDs have allowed many amateur astronomers to make their mark at the leading edge of astronomical research.

Even the most dedicated and successful CCD user would agree, I'm sure, that the products of modern

Fig. 2.4. Urban CCD image of M27, the Dumb-bell Nebula (photo: Nik Szymanek and Ian King).

technology, though they can counter the effects of light pollution for the individual through indirect viewing, do not solve astronomy's problem, and our direct experience of the heavens with the naked eye and with small, more traditional instruments is worth fighting for.

Fig. 2.5. Urban CCD image of M13, the great Globular cluster in Hercules (photo: Nik Szymanek and Ian King).

2.2 Targets

Sample Objects for Light-Polluted Skies

The great majority of amateur astronomers, and just about all of those youngsters across the world who have yet to discover the delights of our oldest science, possess neither filters nor CCD equipment. Chairing a light pollution colloquium in 1993, I made the point that, unless children were "switched on" to the night sky at an early age, not just in science lessons but by direct viewing of the night sky, the number of amateur astronomers in the world would be bound to fall. I asked the audience of about 200 amateurs to raise a hand if they had begun observing, and had got their first telescopes or binoculars, while still of school age. Nearly 200 hands went up.

Today's young amateur astronomer, as well as those with greater experience but modest equipment, need not be discouraged by the lack of pristine skies. Efficient optical aid and some reference material (see bibliography at the end of this book) will reveal hundreds of interesting objects, apart from the Moon and the planets, which can still be enjoyed through moderate light pollution, without filters: double stars of striking colours, bright though neglected clusters, curious starfields, variables, even stars whose motion through our Galaxy you can actually follow ... the list is long.

I offer below a selection of 100 such objects, visible from northern mid-latitudes, seen and recorded beneath a *moderately* light-polluted sky. A general and not very "high-tech" observer, I trawl the night sky, while neighbours sleep, from a small run-off roof observatory in my back garden, in Wimborne, Dorset, south-central England. Wimborne, where my family has lived since 1982, is a small town on the north-western edge of Poole and Bournemouth, which have grown together to form one of the biggest cities on the south coast, with a population of nearly a third of a million: that many people waste a lot of light into space (Fig. 2.6). My sky ranges from slightly light-polluted to the north, away from the neighbouring big town, to severely polluted low in the south, where almost nothing can be seen except on rare, crystal-clear nights.

Fig. 2.6. What a third of a million people throw into the sky: light pollution over Poole and Bournemouth.

I give a brief, and sometimes subjective, description of each object in the list, with observing details. I assure the more experienced observer, who may be already familiar with some of these objects, that there are far more to be found: forty years' observing from back-garden sites in places as different as the East End of London and the under-populated rural county of Shropshire have convinced me that, whether the sky is baleful orange or coal-black, there's always something new to reward the adventurous searcher. I have deliberately included some objects in constellations lying far down in the south from mid-northern latitudes, just to make the point that even down there, it's worth searching on the best nights for the new and unusual in what might be considerable skyglow,

Fig. 2.7. Bob's venerable Charles Frank 21-cm (8.5-inch) reflector.

unless you happen to be looking directly into the glare of a nearby lamp.

When observing the listed objects, I have used, unless otherwise stated, the venerable 21 cm (8.5 inch) Charles Frank f/6 reflector shown in the photo (Fig. 2.7), with its Swift eyepieces of powers ×26 (achromatic Huyghenian), ×50 (Kellner), ×108 (Kellner) and ×323 (orthoscopic), sometimes in tandem with ×2 and ×3 Barlow lenses. Its excellent optics and sturdy construction ensure sharp, steady images rivalling those seen in more modern instruments of similar aperture.

All these observations were made during the years when the all-night lighting (Fig. 2.8) in my street was of an extremely wasteful type. Known familiarly by lighting professionals as "post-tops", the local LPS lights threw, I estimated, about 30–40% of their emissions skywards, and the nearest one to my observatory was 30 metres (about 100 feet) away. It was a great day

Fig. 2.8. The old lamps in my street threw a high percentage of their emissions above the horizontal.

Fig. 2.9. The day the new FCO lamp arrived opposite my observatory.

(with many great nights to follow!) when my local council, whose attitude towards environmentally sensitive lighting became progressively more forward-looking during the 1990s, replaced all the street lights in my area in 1998 with well-directed types, including full-cut-off luminaires (Fig. 2.9) to replace the two post-tops nearest my observing site. The all-night streetlights in my road now shine where their light might conceivably be needed (Fig. 2.10). Dialogue with

Fig. 2.10. The FCO in Fig. 2.9 in action: the light goes where it is needed.

your local lighting engineer can pay, in the generally positive climate which I believe is developing among lighting professionals.

I saw then from my own garden how good-quality lighting works for astronomy: new, fainter stars redrew the constellations (Fig. 2.11), and naked-eye detail – dark rifts and bright projections – became apparent in the once elusive Milky Way; in the south, the Scutum starcloud, several degrees below the celestial equator, became occasionally visible on really clear, calm nights above the more distant glow from nearby Poole, a coastal town of more than 100 000 people. This last phenomenon suggests to me that merely localised better lighting can cause considerable sky improvement in all directions, even if the whole city hasn't been relit.

Fig. 2.11. Better lighting, more stars: looking north from my back garden above two FCO streetlights.

After each constellation name in the selection below is the best time of year to see the suggested objects at their highest in the late evening. This list is by no means exhaustive, and it will, I hope, inspire the reader who observes under indifferent skies to take a fresh look at something not observed for years, or extend the hunt for more targets. They are there!

2.3 One Hundred Objects to Look for in Moderately Light-Polluted Skies

Andromeda (October)

1	γ Andromedae (Almach)	RA 02h 03m	Dec +42° 20′

A superb double star for any size of telescope. The primary (mag. 2.3) is golden yellow. The secondary (mag. 5.1), 9.8" distant, is often listed in handbooks as blue, but it remains resolutely *green* whenever I look at it. Is this just an *idée fixe*, a contrast effect, or a real colour? Try a moderate magnification. I see it well at ×108 (Fig. 2.12).

2	NGC 752	RA 01h 58m	Dec +37° 41′

A scattered cluster of stars, 15 of them above tenth magnitude, filling one square degree. The cluster contains a fine triple star and some pairs. NGC 752 is about 1300 l.y. away. The apparently brightest member (mag. 7.1) may be a foreground object, not associated with the cluster. Best seen in binoculars or through a wide-field, low-power eyepiece, using γ And as a guide: NGC 752 is 5° south of the bright star. Do you see any reddish stars in the northern part of the cluster (Fig. 2.13)?

Fig. 2.12. γ Andromedae is the bright star here beneath Comet C/1995 O1 (Hale-Bopp) on 1997 March 31. The comet's tail sweeps towards the 'W' of Cassiopeia.

Aquarius (August–September)

3	Σ2838	RA 21h 55m	Dec –03° 20′
Halfway between α and β Aquarii, this easy double star, with its yellow primary (mag. 6.3) and bluish secondary (mag. 8.8), lies near what Victorian observer Reverend T.W.Webb called "a curious and beautiful stream of small stars north preceding". A small telescope at moderate power will split this pair, separation 16". The Reverend was right about the small stars.			

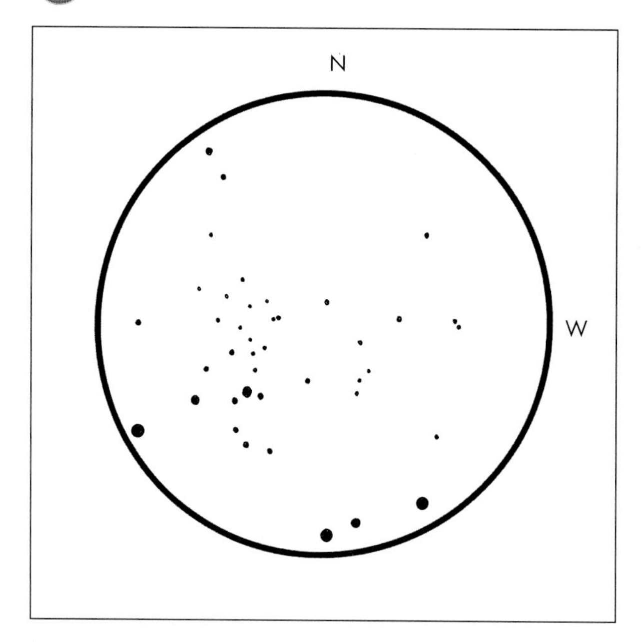

Fig. 2.13. NGC 752.

Object: NGC 752
Constellation: Andromeda
Type: Open cluster
Magnitude: 6
Number of stars: 70+
***Norton's 2000.0* chart:** 3
***Uranometria 2000.0* chart:** 92
Instrument: 6-cm/2.5-in refractor
Magnification: ×30
Field diameter: 45′
Seeing: Ant. II
Light pollution: moderate
Date: 1976 November 19
Time: 2300 UT
Notes: Very large group, best in binoculars, with a fine bright triple and some interesting pairs.

4	τ^1 Aquarii	RA 22h 48m	Dec −14° 03′

Primary mag. 5.8, secondary 9. This pair, 23.7" apart and easily separated with a moderate power, is an optical double. Unremarkable colours, unlike nearby τ^2, a beautiful mag. 5 orange star with a faint blue line-of-sight companion preceding.

5	

Close by τ^1 Aquarii, 4° to the north-east, is a remarkable chain of mag. 8 stars containing the double star Σ2970 (R.A. 23h 02m Dec. −11° 20′; mags. 8.5, 9, sep. 8.5"). Low to moderate powers. Sweep around (Fig. 2.14)!

Aquila (July)

6	57 Aquilae (Σ2594)	RA 19h 54m	Dec −08° 10′

What colour is the secondary star of this duo, easily split at ×40? Nineteenth-century observers, happy to record their subjective impressions, saw the mag. 6.2 companion as "lilac" (Webb), "pale blue" (Knott), and "azure white" (Dembowski). The yellow primary, mag. 5.2, is 36" away.

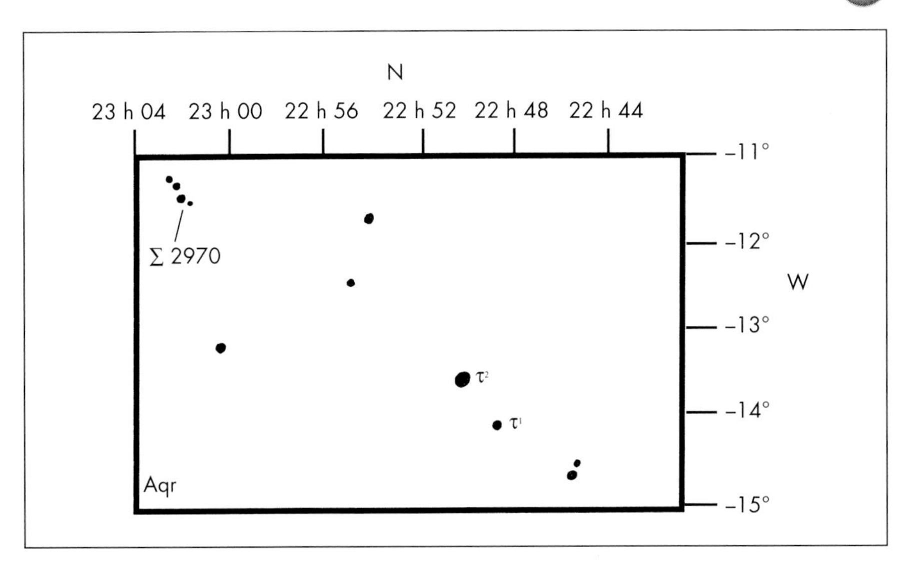

Fig. 2.14. τ^1 Aqr to Σ2970.

Object: Σ2970
Constellation: Aquarius
Type: Double star
Magnitudes: 8.5, 9
Separation: 8.5″
***Norton's 2000.0* chart:** 4
***Uranometria 2000.0* chart:** 303

7	R Aquilae	RA 19h 06m	Dec +08° 14′

Many variable stars approach naked-eye visibility, but do not share the renown of those which, like Algol and Mira, have bright maxima. Long-period variable (LPV) R Aql is a fine red star, easily found with binoculars when at its maximum magnitude of 5.5. Use Altair and ζ Aquilae to locate it (see chart). For how long can you follow it, as it dims towards its minimum of mag. 12? Its period of variability is 284 days, a value which is slowly decreasing (a detailed examination by J. Greaves and J. Howarth of this star's behaviour appeared in the *Journal of the British Astronomical Association*, June 2000, pp 131–42). Predictions for maxima/minima of many variable stars can be found in popular astronomy magazines such as *Sky & Telescope* or *Astronomy Now*, or websites such as www.telf-ast.demon.co.uk/ (BAA Variable Star Section) (Fig. 2.15).

Aries (October–November)

8	1 Arietis (Σ174)	RA 01h 50m	Dec +22° 17′

Nearly 2° north of Sheratan (β Arietis), therefore easily found, and bright enough (mags. 6 and 7) for a small telescope. A fairly high magnification (×100+) will reveal a fine golden-yellow and blue pair, 2.8" apart.

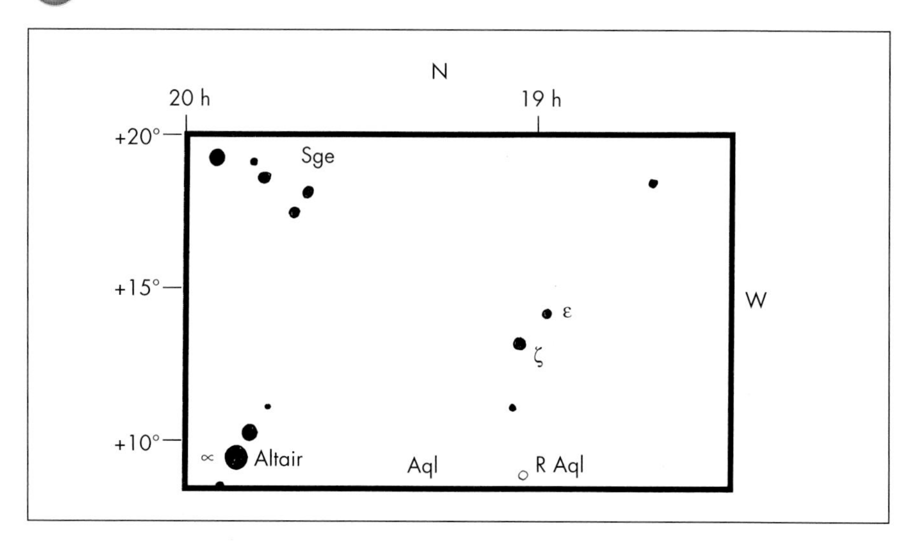

9	λ Arietis (OΣΣ21)	RA 01h 58m	Dec +23° 35′

A fine wide double for a small telescope, mags. 4.9 and 7.7, sep. 38.5". Is the primary white or blue? Easily found 2° west of α Ari (Hamal).

Fig. 2.15. Finding R Aql.

Object: R Aquilae
Type: Long-period variable
Period: 284 days
***Norton's 2000.0* chart:** 13
***Uranometria 2000.0* chart:** 206

Auriga (December)

10	NGC 1907	RA 05H 28m	Dec +35° 20′

An interesting compact cluster, close to the Galactic Equator, and half a degree south of the bright cluster M38, which serves as a signpost. NGC 1907 bears high powers well, and many fainter cluster members suggest themselves at the limit of visibility. Averted vision can help bring out these tantalisingly faint stars, as it can with many of the items in this list. I find that averted vision is often more effective if the field is allowed to drift and the eye concentrates on a fixed point, rather than moving it to "hunt" for the right distance from the object (Fig. 2.16).

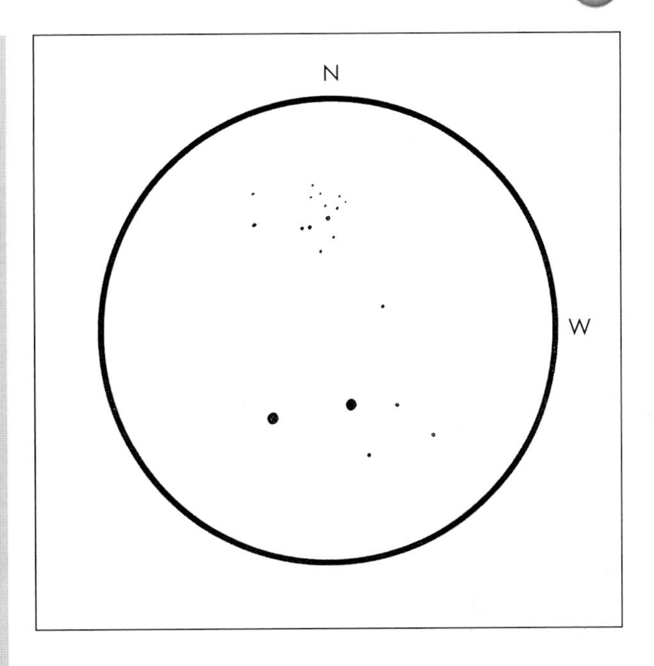

Fig. 2.16. NGC 1907.

Object: NGC 1907
Constellation: Auriga
Type: Open cluster
Magnitude: 8
Number of stars: 40
***Norton's 2000.0* chart:** 5
***Uranometria 2000.0* chart:** 97
Instrument: 21-cm./8.5-in. reflector f/6
Magnification: ×324
Field diameter: 8′
Seeing: Ant. III
Light pollution: moderate
Date: 1989 April 7
Time: 2130 UT

Notes: Many stars of mags. 10+. 'Haze' of many members just below limit of visibility.

11	14 Aurigae (Σ653)	RA 05h 15m	Dec +32° 45′

Centred in an interesting field at ×50, the two components of 14 Aur show a marked colour contrast: pale yellow and white (though I once recorded the secondary as "greenish"). The primary is slightly variable (KW Aur, described in *Sky & Telescope* December 1990, p. 610). Mags. 5.1–5.2, 7, sep. 15". There is a faint mag. 11 companion 10" to the north (Fig. 2.17).

Boötes (April–May)

12 13	ι and κ Boötis	RA 14h 15m	Dec +51° 30′ (centre of field)

This neat pair of pairs fits into a medium-power field. You will need a field diameter of at least 45′. All four stars look slightly different in hue – do you agree? Data: ι, mags. 5.3, 7.5 (variable?), sep. 2.2"; κ, mags. 4.6, 6.6, sep. 13.4".

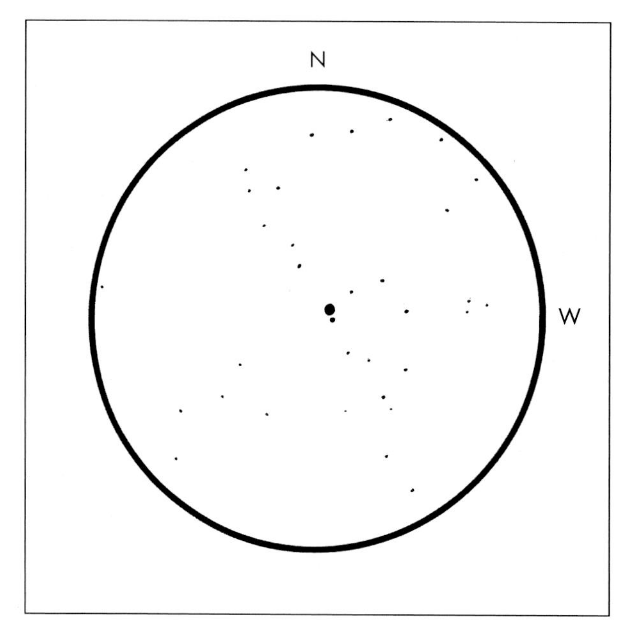

Fig. 2.17. Field of 14 Aur.

Object: 14 Aurigae
Type: Double star
Magnitudes: 5.1–5.2, 7.9
Separation: 15″
***Norton's 2000.0* chart:** 5
***Uranometria 2000.0* chart:** 97
Instrument: 21-cm./8.5-in. reflector f/6
Magnification: ×50
Field diameter: 45′
Seeing: Ant. III
Light pollution: moderate
Date: 1990 March 25
Time: 2229 UT

Notes: A fine colour contrast in a crowded field.

Camelopardus (December)

14	Kemble's Cascade (The Wristwatch)	RA 04h	Dec +63°

I happened upon this fascinating binocular object quite by chance many years ago, and let out a whistle of disbelief. A wristwatch, or pendant necklace, opened out in the sky! Straddling four 50′ telescope fields, this delicate chain of stars, seen in the photo (Fig. 2.18) in the same field as Comet Hyakutake, nudges with its southern tip the dainty little telescopic cluster NGC 1502. Of all the many star chains and chance alignments in the sky, the Cascade must take the prize. See *Sky & Telescope*, December 1980, p. 547, for more information.

15	Chain of pairs, Camelopardus –Cassiopeia	RA 03h 12m to 32m,	Dec +67° to +68°

The sketch (Fig. 2.19) shows, in three overlapping fields, a curious sequence of five double stars, all mags. 7/8, straddling the Camelopardus–Cassiopeia border, inviting some interesting exploring with both low and high powers. The pairs are: OΣ54 Cam (mags. 7, 8.5, sep. 23.6", p.a. 358°); Σ374 Cam (mags. 7, 8.5, sep. 10.9", p.a. 295°); Hu1056 Cam (mags. 8, 8, sep. 1", p.a. 269°); SAO 12680/-81 Cas (my estimates: mags. 7, 9, sep. 25", p.a. 175°); unidentified pair (Cas) (my estimates: mags. 8.5, 8.5, sep. 10", p.a. 195°).

Fig. 2.18. Kemble's Cascade ("The Wristwatch") to the left of Comet C/1996 B2 (Hyakutake), 1996 March 30.

Cancer (January–February)

16	57 Cancri	RA 08h 54m	Dec +30° 40′

A ragged line of stars leads north-east from the colourful mag. 4 double star ι Cancri towards 57, two degrees away. This close pair (sep. 1.5"), mags. 6 (yellow) and 6.5 (yellow-orange), when closely examined, turns out to be a triple: an elusive 9th-mag. companion lies 50" away to the south. Use high powers.

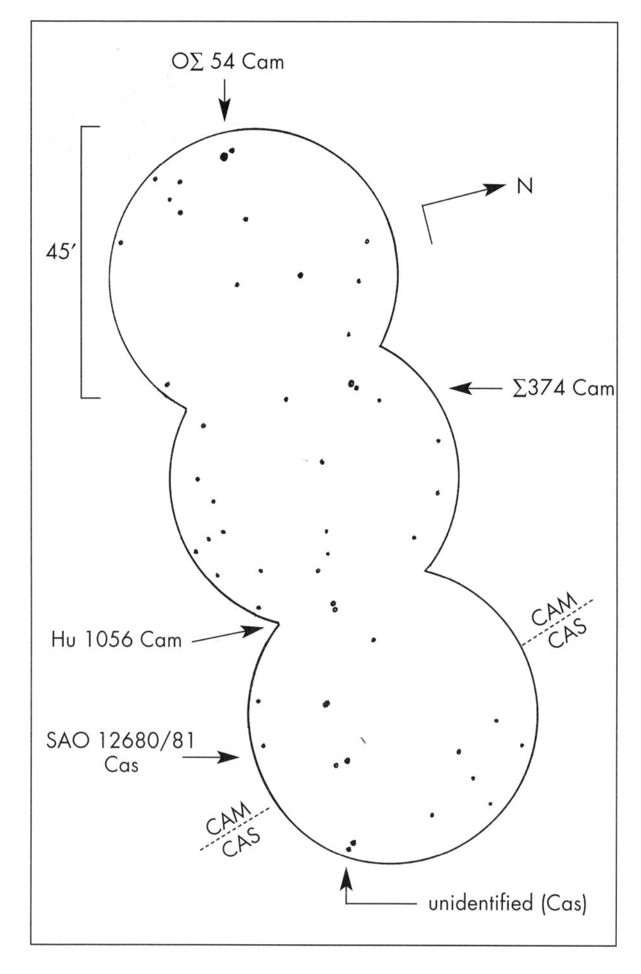

Fig. 2.19. Chain of pairs in Camelopardus and Cassiopeia.

***Norton's 2000.0* chart:** 2
***Uranometria 2000.0* chart:** 18
Instrument: 21-cm. /8.5-in. reflector f/6
Magnification: ×50
Field diameter: 45′ (three fields)
Ant. II
Light pollution: moderate
Date: 1990 October 12
Time: 2250 UT

Notes: Some interesting sweeping around this chain of pairs.

17	M67 (NGC 2682)	RA 08h 50m	Dec +11° 49′

M67 is a good example of a "neglected neighbour", an object not much discussed and often forgotten about because it is upstaged by a showier target nearby: in this case, the Beehive cluster, M44, nine degrees to the north. *Norton's 2000.0* calls M67 "rich", though it is fairly compact at only 15¢ across. Of its 500 true members, none is brighter than mag. +10. A moderate telescope and low power will show a dozen stars in a straggling line, with a hint of many more just beyond the limit of visibility in a not too polluted sky (Fig. 2.20).

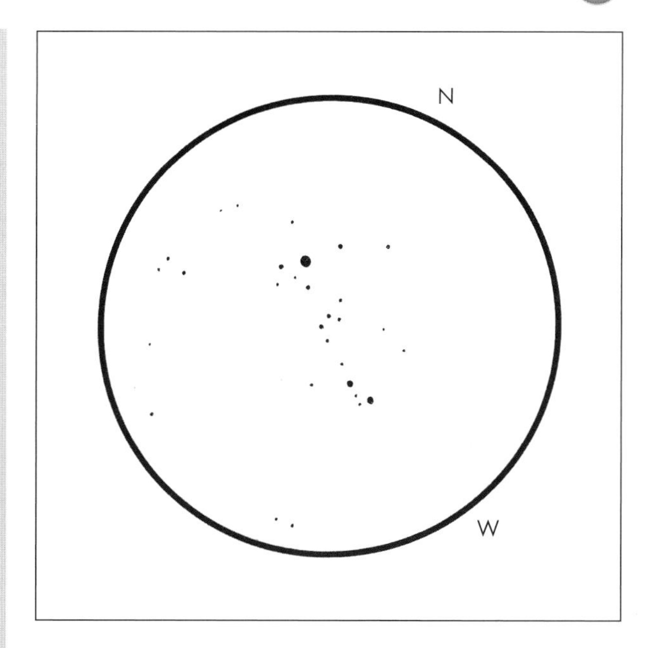

Fig. 2.20. M67.

Object: M67 (NGC 2682)
Constellation: Cancer
Type: Open cluster
Magnitude: 7
Number of stars: 500+
***Norton's 2000.0* chart:** 7
***Uranometria 2000.0* chart:** 187
Instrument: 6-cm./2.5-in. refractor
Magnification: ×30
Field diameter: 45′
Seeing: Ant. II
Light pollution: moderate
Date: 1990 January 30
Time: 2200 UT

Notes: Compact, with many faint stars at limit of visibility.

Canes Venatici (April)

18	Y Canum Venaticorum ("La Superba")	RA 12h 45m	Dec +45° 26′

Rightly called the "superb" star by Secchi, Y CVn is not only a glorious red colour, but also one of the brightest stars in this fairly dim constellation. Easily found, as it forms a right angle with α CVn and β CVn (Fig. 2.21), Y takes 157 days to fall from mag. 5.2 to mag. 6.6 and rise again. One of the most intriguing telescopic gems you can show a non-astronomer neighbour who thinks all stars are white is a relatively bright star of vivid red hue, such as Y CVn, R Lep ("Hind's Crimson Star") or μ Cep ("The Garnet"). The diameter of Y CVn, a cool (~2600° C) red giant, is about 400 million kilometres/250 million miles: if this star replaced the Sun, the Earth's orbit would lie about 50 million kilometres/30 million miles inside the star's "surface".

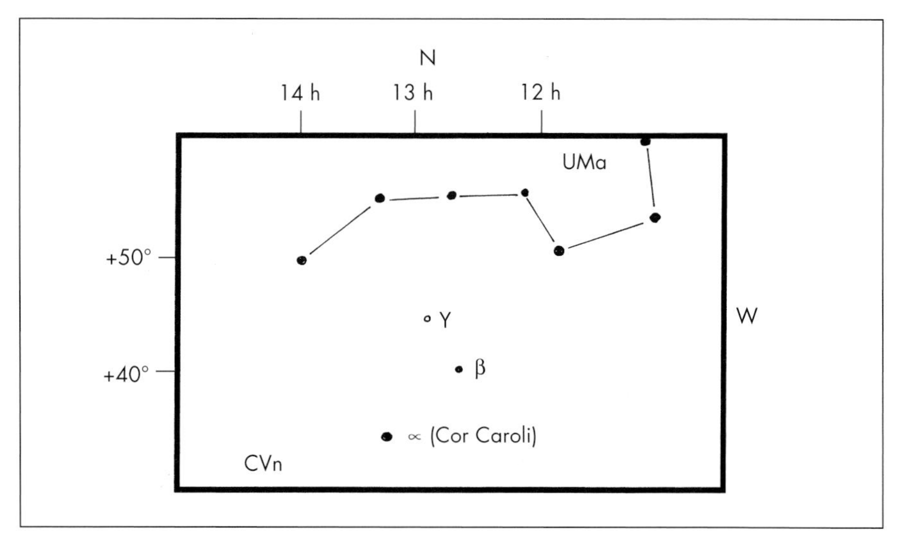

Fig. 2.21. Finder chart for Y Canum Venaticorum.

Object: Y Canum Venaticorum ('La Superba')
Type: Semi-regular variable
Period: ~160 days
***Norton's 2000.0* chart:** 9
***Uranometria 2000.0* chart:** 75

19	M3 (NGC 5272)	RA 13h 42m	Dec +28° 23′

You may not see this 6th mag., slightly oval globular cluster quite as well as the Reverend Webb saw it in the better skies of the 1870s, "blazing into a confused brilliancy towards the centre, with many outliers", but it is certainly a fine sight with moderate powers and a modest 10-cm/4-in telescope. The outlying stars seem to form streams. This starry mass, 40 000 l.y. away, lacks the hint of darker patches seen in the central region of the better known globular M13. M3 contains about half a million stars in a spherical space about 220 l.y. across.

Canis Major (December –January)

20	h3945	RA 7h 17m	Dec –23° 18′

Why doesn't my *Norton's 2000.0* describe in its listings h3945 (or 145 CMa), which shows the finest colour contrast of any double star in the winter sky? Look for this pair, orange-red and electric blue, just east of o^2 CMa and north of τ. Finding it by chance for the first time in 1971, low in the south through the light pollution of Poole, I was amazed that I had never heard of this celestial gem before. A must! Mags. 4.8, 6; separation 27", easily split with moderate power. This is an optical double, only a line-of-sight effect; the brighter of the two stars is 2500 l.y. from us, while its "neighbour" is only one-tenth of that distance away. The finder chart (Fig. 2.22) does not include all the stars in this rich area (h3945 is just within the traditional Milky Way "boundary"), and is meant for "star-hopping" using δ and τ as pointers.

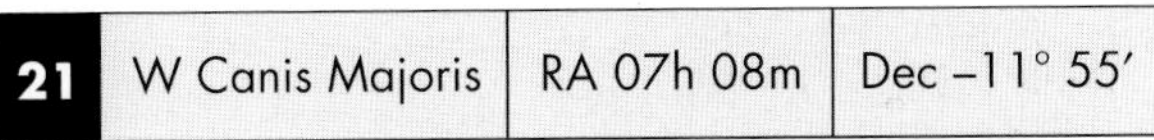

21	W Canis Majoris	RA 07h 08m	Dec –11° 55′

A slow irregular variable, close to the Galactic Equator and worth watching. Its maximum magnitude (6.35) makes it an easy binocular object. One of the N-spectrum "carbon stars", characteristically red . Minimum mag. 7.9.

Fig. 2.22. A "star-hop" to h3945.

Object: h3945
Constellation: Canis Major
Type: Double star
Magnitudes: 4.8, 7
Separation: 27″
***Norton's 2000.0* chart:** 8
***Uranometria 2000.0* chart:** 319

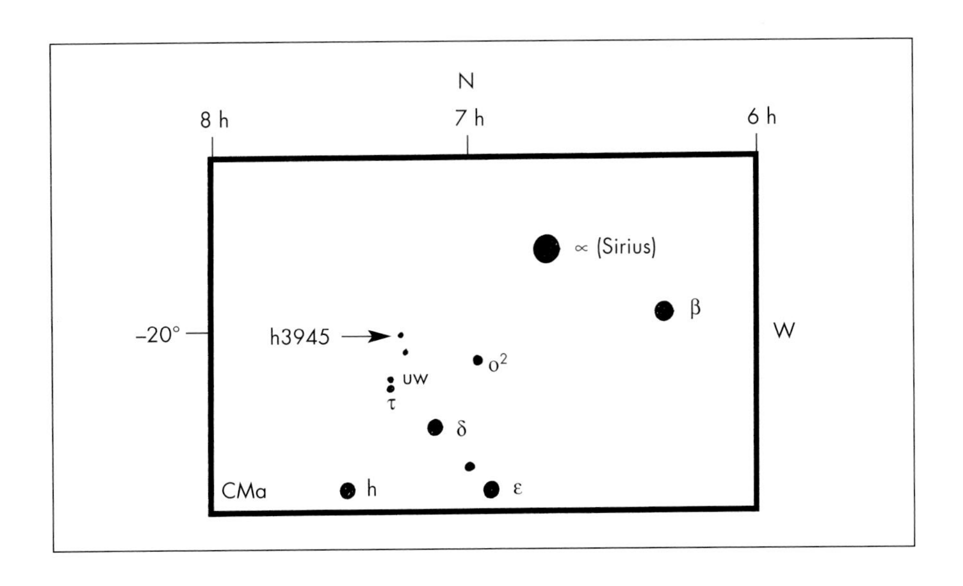

Canis Minor (January–February)

22	Σ1149	RA 07h 49m	Dec +03° 12′

A double star, low in the winter sky for northern observers, but well worth the hunt. At the western end of a line of mag. 6 and 7 stars, some themselves double, strung out just south-east of Procyon, this pair is yellow and blue, mags. 7.5 and 9, 22" apart, position angle 41°. A medium power on a small instrument, e.g. 7.5 -cm/ 3 -inch, will split it.

23

More of a challenge at the other end of the line of stars indicating Σ1149 is OΣ182, a brighter pair (7, 7.5), but only 1" apart. Highest power needed (Fig. 2.23).

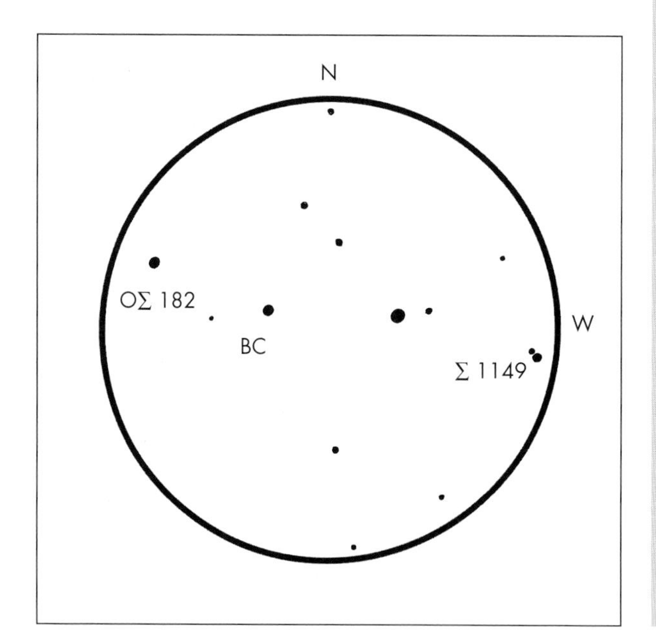

Fig. 2.23. Stars near Σ1149.

Object: Σ1149
Constellation: Canis Minor
Type: Double star
Magnitudes: 7.5, 9
Separation: 22″
***Norton's 2000.0* chart:** 7
***Uranometria 2000.0* chart:** 230
Instrument: 21-cm./ 8.5-in. reflector f/6
Magnification: ×32
Field diameter: ~1°
Seeing: Ant. II
Light pollution: moderate
Date: 1989 April 2
Time: 2030 UT

Notes: Striking 'snake' of four main field stars.

Capricornus (August)

Down in the worst of the light pollution for many observers, Capricornus still offers some interest.

24	α Capricorni	RA 20h 17m	Dec –12° 30′

Algiedi ("The Goat") is a wide naked-eye pair, 380" apart, and splendid in a low-power field. In reality, however, α^1 (700 l.y.) is more than six times further from us than α^2 (108 l.y.), the more southerly and brighter of the two. α^1, mag. 4.5, has a difficult 9th mag. companion 45" away, while α^2, mag. 3.5, is also double, the companion being of mag. 11 at 6.6".

Cassiopeia (October)

High above in autumn, in what is often the least polluted part of the sky, Cassiopeia is rich in worthwhile objects.

25	OΣΣ254 (primary is WZ Cassiopeiae)	RA 00h 02m	Dec +60° 25′

A fine wide double with good colour contrast. Deep orange-red (mag. 7) and striking blue (mag. 8), well seen with a medium-sized reflector (15–21 cm/6–8 inch) at about ×100. Separation 58". In a crowded field (Fig. 2.24), and the colours are superb when the red semi-regular variable WZ is at its brightest (magnitude range 7–10, period 186 days).

26	NGC 457	RA 01h 19m	Dec +58° 20′

One of the most pleasingly arranged clusters in the sky, and easily seen even with low power (e.g. ×32) and a modest telescope. Highly luminous supergiant ϕ Cas (mag. 5) is involved, making it easy to find. My daughter, seeing it for the first time, exclaimed: "It's a little stickman with two big, bright eyes!" With south upwards, as shown by the chart (Fig. 2.25), the cruciform cluster suggests just this at low powers. Sometimes called the "Owl" cluster (which may lead to some confusion with planetary nebula M97, the Owl Nebula), but the stickman can't be ignored. Distance about 8000 l.y. Recent *Hipparcos* measurements suggest that the two "eyes", ϕ Cas and HD7902, are foreground stars, between 2000 and 3000 l.y. away.

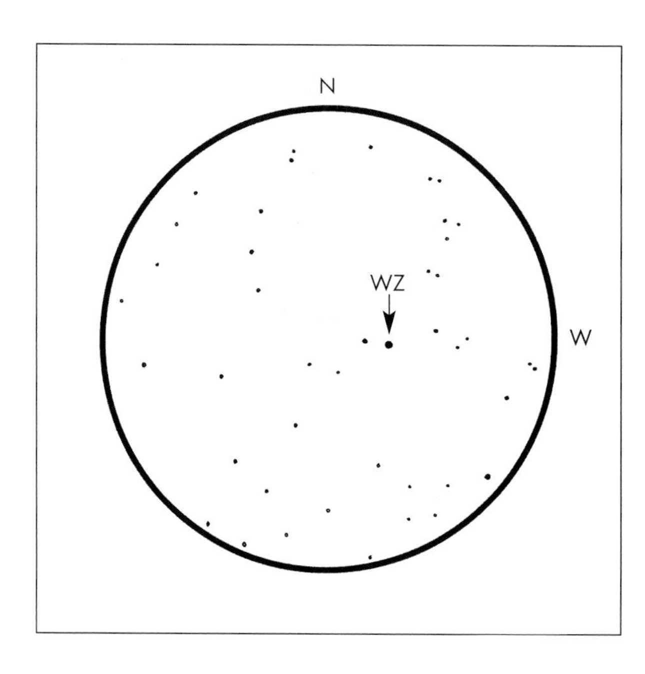

Fig. 2.24. Field of WZ Cas.

Object: OΣΣ 254 (primary is WZ Cas)
Constellation: Cassiopeia
Type: Double star
Magnitudes: 7, 8
Separation: 58″
***Norton's 2000.0* chart:** 2
***Uranometria 2000.0* chart:** 35
Instrument: 21-cm./8.5-in. reflector f/6
Magnification: ×108
Field diameter: 21′
Seeing: Ant. I
Light pollution: moderate
Date: 1990 September 17
Time: 2215 UT

Notes: Colours very striking: deep orange-red, electric blue.

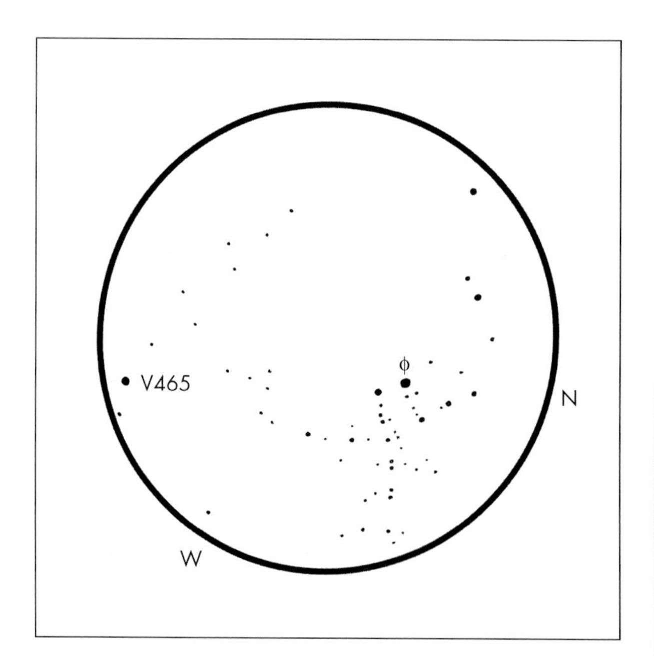

Fig. 2.25. NGC 457.

Object: NGC 457
Constellation: Cassiopeia
Type: Open Cluster
Magnitude: 6.5
Number of stars: ~100
***Norton's 2000.0* chart:** 2
***Uranometria 2000.0* chart:** 36
Instrument: 21-cm./8.5-in. reflector f/6
Magnification: ×32, ×108
Field diameter: 50′
Seeing: Ant. I
Light pollution: slight
Date: 2000 July 29
Time: 2229 UT

Notes: Grand, cruciform group. ϕ yellow, fine contrast with blue neighbour mag. 7 and V465 (red).

27	Trumpler 1	RA 01h 36m	Dec +61° 18′

Between two of Cassiopeia's showpiece clusters, M103 and NGC 663, is a modest little group of about 25 stars, the brightest of them about mag. 10. What makes Tr 1 worth a look at high powers (×100+) is the remarkable north–south line-up of the four main stars, an intriguing sight at ×323. Easily found by moving ahead of NGC 663 in R.A.

Cepheus (August–September)

(28, 29, 30) An interesting find: three multiple stars in the same ×50 field (diameter 45′), in a crowded area. Try centring your instrument on: Σ2816, at R.A. 21h 39m, Dec. +57° 30′.

The chart (Fig. 2.26) shows three members of this quadruple system, with Σ2813 and Σ2819 bracketing it. Data:
Σ2813: mags. 8.5, 9, sep. 10″;
Σ2816: mags. 6.3, 7.9, 8, 13, seps. 11″, 20″, 1.6″;
Σ2819: mags. 7.5, 8.5, sep. 12″.

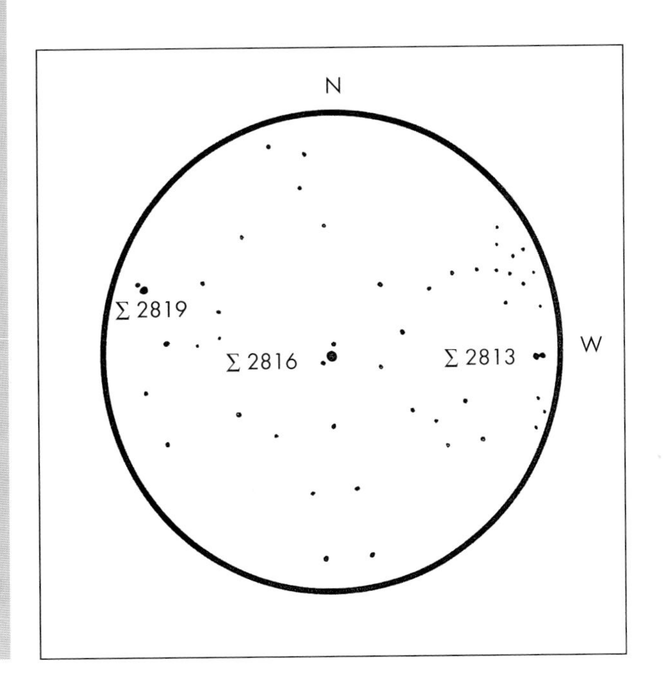

Fig. 2.26. Σ2813, Σ2816, Σ2819.

Objects: Σ2813/2816/2819
Constellation: Cepheus
Type: Double stars
Magnitudes: see text
Separation: see text
***Norton's 2000.0* chart:** 2
***Uranometria 2000.0* chart:** 57
Instrument: 21-cm./8.5-in. reflector f/6
Magnification: ×50
Field diameter: 45′
Seeing: Ant. II–III
Light pollution: moderate
Date: 1991 August 14
Time: 2330 UT
Notes: Three intriguing pairs chase slowly across a superb field.

31	RW Cephei	RA 22h 23m	Dec +55° 58′

Not a bright star, this irregular variable, but if you can catch it near the top of its range (mag. 7.6, falling to 9), do you see it as pink? Use ε Cep as a starting point for your "star-hop" towards RW's location by the border with Lacerta: stars down to mag. 8 are shown only for the preceding side of the chart (Fig. 2.27).

32 33	NGC 6939 NGC 6946	RA 20h 31m RA 20h 35m	Dec +60° 40′ Dec +60° 09′

The old cliché "stardust" comes to mind as open cluster 6939 slides into a medium-power field. A mass of fairly faint stars packs its centre, and averted vision suggests many more, frustratingly at the limit of perception. Sc galaxy 6946, resembling a shrunken M33 in photographs, and exactly straddling the Cepheus–Cygnus border, appears as a small, ghostly, fleetingly glimpsed disc in the same field as 6939 at ×50. There is a splendid photo showing both objects in *Burnham's Celestial Handbook*, p. 620. The galaxy is also noteworthy for relatively frequent discoveries of supernovae (1917, 1939, 1948, 1968, 1969, 1980), which is probably related to the fact that it is only about 15 million l.y. from the Milky Way. Worth a frequent look, therefore.

Fig. 2.27. A "star-hop" to RW Cep.

Object: RW Cephei
Type: Irregular variable
***Norton's 2000.0* chart:** 2
***Uranometria 2000.0* chart:** 57

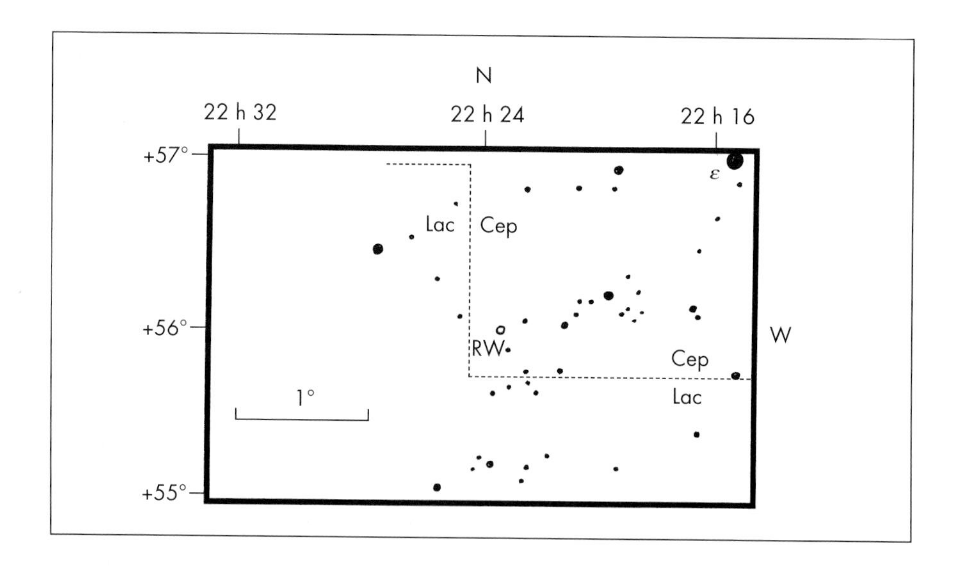

34	NGC 7538	RA 23h 14m	Dec +61° 30′

A superb colour photo of this complex of nebulae on the Cepheus–Cassiopeia border appeared on the cover of *Sky & Telescope* in August 1991 (though it was wrongly assigned to Perseus). I looked for it that same month, on a very clear and still night, at ×108, and was delighted to see not only a hint of the three main areas of nebulosity, but some internal brightness gradation. I have never found it again. Definitely a challenge for the most transparent nights!

Cetus (October–November)

Though low in the sky for observers in northern mid-latitudes, Cetus offers:

35 36	37 Ceti and Σ101.	RA 01h 14m	Dec –7° 55′ (37 Cet)

In the same low-power field, this pair of pairs is a striking sight, if challenging because of their low declination, with possible strong light pollution. Data: 37 Cet: mags. 5, 7, sep. 50”; Σ101: mags. 8, 10, sep. 21”, located just west of ϑ Ceti (mag. 3.8). Webb described both pairs as “yellow and orange”.

Coma Berenices (March–April)

37	Melotte 111	RA 12h 24m	Dec +26° (centre of field)

To the unaided eye, the constellation of Berenice’s Hair is a vague and inconspicuous scattering of sub-4th magnitude stars. Several of its brightest members belong to the nearby galactic cluster Melotte 111, about 250 l.y. distant. Tainted skies may not allow you to share G.P.Serviss’ poetic impression of this group (“gossamer spangled with dewdrops”), but try Mel 111 with low-power binoculars when it is high.

38	Σ1639	RA 12h 24m	Dec +25° 35′

Try a high power on this one: at ×323, I opened up the 1.3” gap between these two pale yellow stars, mags. 6.6, 7.8. In the middle of Mel 111 (see above), and one degree preceding and slightly south of the wide double 17 Com.

Corona Borealis (May)

39	T Coronae Borealis	RA 15h 59m	Dec +25° 55′

This one is for optimists only! Keep a regular binocular watch on the area just below the little Crown (Fig. 2.28), where lurks a recurrent nova: T CrB, the "Blaze Star", a variable star normally at mag. 10.3, which *might* just have a surprise in store. The larger component of T CrB is thought to be a stable red giant, with a smaller blue companion which exhibits unpredictable fluctuations due to material falling upon it from its bloated neighbour. A fierce 1866 outburst reached mag. 2. Some 80 years later, in February 1946, T CrB brightened to mag. 3, its outer layers bursting away at an estimated 2700 miles (4320 kilometres) per second, according to Burnham. Another 80 years would take us to 2026, but we may not have to wait until then for another paroxysm: John Herschel noted a brightening in 1842 to mag. 6, just 24 years before T CrB's 1866 display.

40	ζ Coronae Borealis	RA 15h 39m	Dec +36° 38′

An easy double (mags. 5, 6, sep. 6.3") for a small telescope and moderate power. A beautiful white-white pair. Webb calls the stars "greenish".

Fig. 2.28. Finder chart for T CrB.

Object: T Coronae Borealis
Type: Recurrent nova
***Norton's 2000.0* chart:** 11
***Uranometria 2000.0* chart:** 155

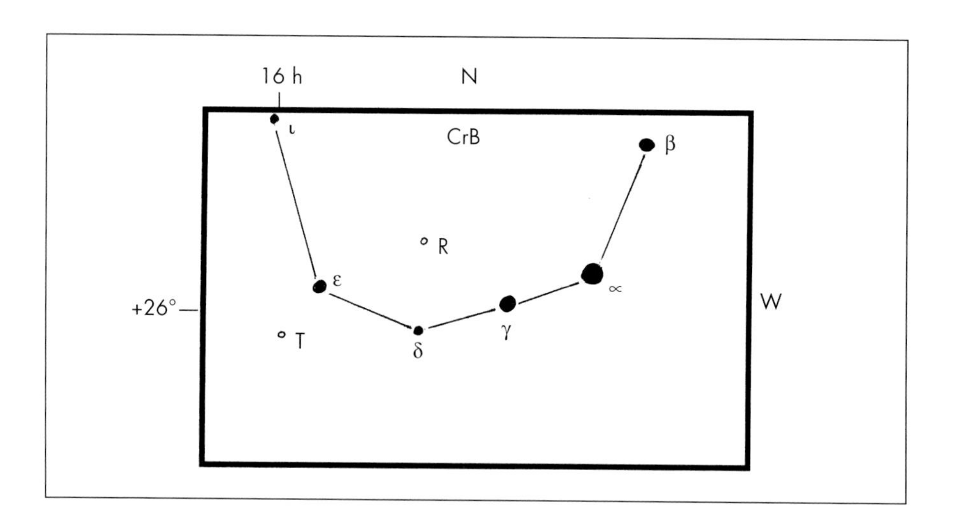

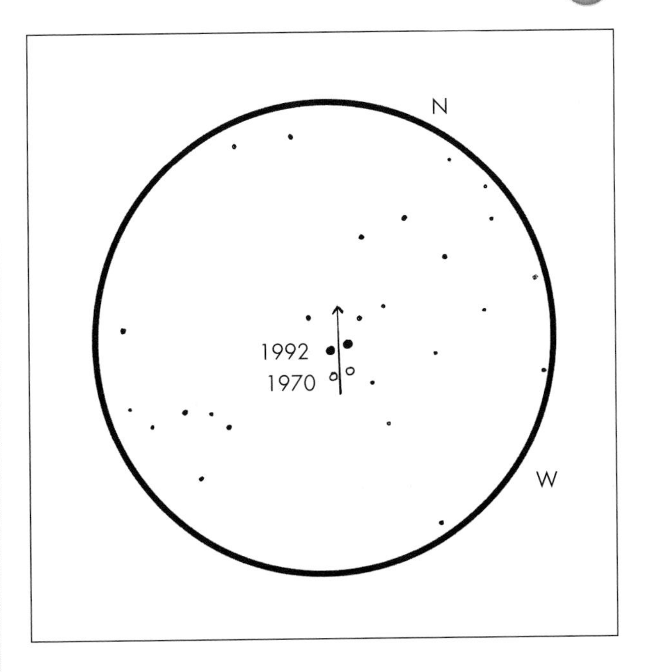

Fig. 2.29. Motion of 61 Cyg: positions in 1970 and 1992.

Object: 61 Cygni
Type: Double star
Magnitudes: 5.3, 5.9
Separation: 30″
***Norton's 2000.0* chart:** 13
***Uranometria 2000.0* chart:** 121
Instrument: 21-cm./8.5-in. reflector f/6
Magnification: ×108
Field diameter: 21′
Seeing: Ant. II
Light pollution: moderate
Date: 1970 July 10, 1992 July 31
Time: 2300 UT, 2318 UT

Notes: Orange pair in a fairly crowded field.

Cygnus (July–August)

41	61 Cygni	RA 21h 07m	Dec +38° 45′

See the Galaxy rotate! Well, not quite, but 61 Cyg ("Piazzi's Flying Star") is a binary star whose motions you can measure and record over the years with a small telescope. The two components, mags. 5.3 and 5.9, both orange, perform a slow 700-year orbital dance, but of more interest is the abnormally rapid annual proper motion of 5.22" north-eastwards against the background stars, due to 61 Cyg "passing" the Sun at a distance of only eleven l.y. The sketch (Fig. 2.29) shows my observations of its positions in 1970 with open circles, and in 1992 as filled circles. Where do you see it now?

42	Region of SAO 50246	RA 20h 55m	Dec +47° 25′

This unremarkable 6th-magnitude star is preceded by a marvellous field of fainter stars in curious chains and swirls (Fig. 2.30). Low power and a wide field reveal the whole grand sight. *Uranometria 2000.0* shows nebula I.5076 around the star: I see no sign of it, but can the star chains be related?

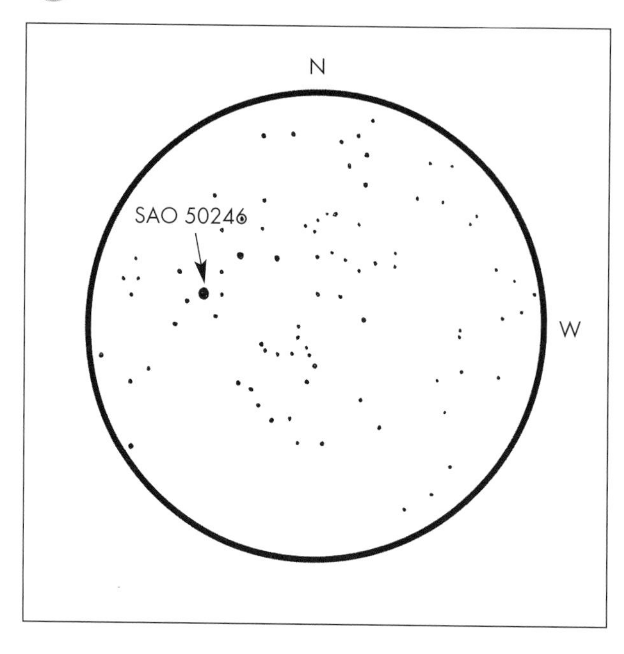

Fig. 2.30. Star chains near SAO 50246.

Objects: Field of SAO 50246
Constellation: Cygnus
***Norton's 2000.0* chart:** 13
***Uranometria 2000.0* chart:** 85
Instrument: 21-cm./8.5-in. reflector f/6
Magnification: ×50
Field diameter: 45′
Seeing: Ant. II
Light pollution: moderate
Date: 1992 July 28
Time: 2255 UT

Notes: Splendid lanes of faint stars. Not listed as a cluster.

43	NGC 6866	RA 20h 04m	Dec +44° 10′

A large and beautiful open cluster, with many faint stars in branching lines, suggesting open wings. About 40 stars seen at ×79. Find 6866 by moving one degree south of δ Cyg (R.A. 19h 45m), and then 19 minutes in increasing R.A.(or nearly five degrees) to 20h 04m.

Delphinus (July–August)

44	NGC 7006	RA 21h 02m	Dec +16° 11′

Not a hard one to pinpoint, but a faint one! If you put γ Delphini, the Dolphin's "nose", at the centre of a medium-power field (say, ×108), and sweep slowly eastwards for nearly four degrees, you might just catch a glimpse of NGC 7006, the globular cluster which holds the record as the furthest such object from us in the Milky Way's halo, at 130 000 l.y. (Burnham) – if you don't include the extra-galactic "wanderer" NGC 2419 in Lynx (~250 000 l.y), described below. NGC 7006 is elusive at mag. 11; I *think* I once saw it fleetingly as a faint smudge, with averted vision, through the 21 cm/8.5 inch reflector at ×216 (×108 with ×2 Barlow).

45	NGC 6934	RA 20h 34m	Dec +07° 20′

No luck with 7006? Console yourself with this small but brighter globular, mag. 9, four degrees almost due south of the Dolphin's "tail" star, ε Delphini. It looks like a slightly out-of-focus star at ×108. Bluish?

Draco (March–June)

46	Σ2398	RA 18h 43m	Dec +59° 45′

A binary with extremely rapid proper motion (2.3" per year), being only 11.3 l.y. away. At intervals of just a few years, you can track its path across the sky against background stars. The sketch (Fig. 2.31) shows an observation of its position in 1967 (cf. *Burnham's Celestial Handbook*, p. 869) with open circles,and my observation of its changed position in 1989 as filled circles. Not shown in *Norton's 2000.0*, but easily found one degree preceding 47 (Omicron) Dra. Mags. 8, 8.5, sep. 15.3", position angle 163°, both red dwarfs. Easy with low power.

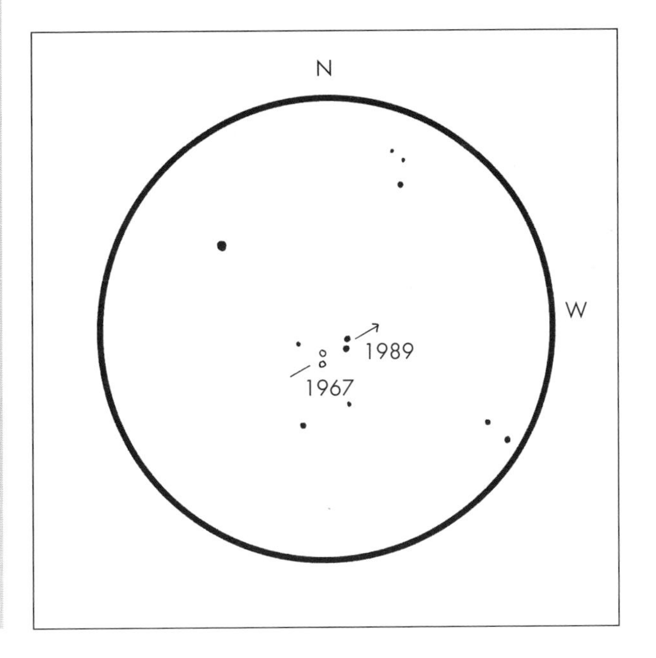

Fig. 2.31. Motion of Σ2398 against background stars: positions in 1967 and 1989.

Object: Σ2398
Constellation: Draco
Type: Double star
Magnitudes: 8, 8.5
Separation: 15″
***Norton's 2000.0* chart:** 2 (1° p. 47 Dra)
***Uranometria 2000.0* chart:** 54
Instrument: 21-cm./8.5-in. reflector f/6
Magnification: ×100
Field diameter: 25′
Seeing: Ant. II–III
Light pollution: moderate
Date: 1989 June 18
Time: 2334 UT

Notes: Motion in 22 years approximately 50″.

47	5, 6 Draconis and SAO 7611	RA 12h 36m	Dec. +70° (centre of field)

A bright trio, mags. 4, 5 and 7, easily seen in binoculars. Aim a telescope at them, with a low power for a wider field, and admire the colours. I see them as, respectively, blue-white, orange and yellow. There is a curious echo of this set of stars two degrees away to the north: a fainter trio (mags. 8, 8, 9) in the same pattern and orientation.

48	Field of stars centred on	RA 18h 35m	Dec +72° 25′

A miniature mirror-image of the main stars of Cassiopeia! A pleasingly symmetrical group of mag. 6 to mag. 8 stars half a degree across, but I can't find them listed as a cluster. Low power best (Fig. 2.32).

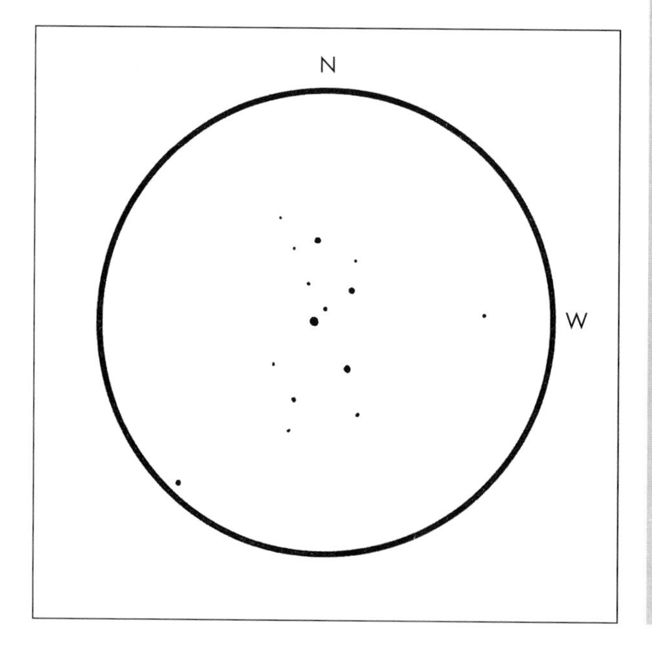

Fig. 2.32. A miniature Cassiopeia in Draco.

Objects: Starfield centred on R.A. 18h 35m, Dec. +72° 25′
Constellation: Draco
Magnitudes: 6-8
***Norton's 2000.0* chart:** not shown
***Uranometria 2000.0* chart:** 12
Instrument: 21-cm./8.5-in. reflector f/6
Magnification: ×50
Field diameter: 45′
Seeing: Ant. III
Light pollution: moderate
Date: 1989 June 20
Time: 2329 UT

Notes: A pleasingly balanced group, immediately suggesting a mirror image of Cassiopeia.

Gemini (December–January)

49	α Geminorum (Castor)	RA 07h 35	Dec +31° 53′

Not hard to find! This famous sextuple system consists of three pairs, Castor A, B and C, 52 l.y. away.
Castor A: mag. 1.9;
Castor B: mag. 2.9, A-B sep. 3.9";
Castor C [YY Gem], variable, mag. 8.9-9.6, 70" away to the south.
It was the rapid 400-year orbit of Castor B around Castor A which finally convinced William Herschel of the reality of binary systems. Webb saw A as "greenish". Try highest magnifications.

50	δ Geminorum (Wasat)	RA 07h 20m	Dec +21° 59′

Although δ is in itself interesting, being a fine binary (mags. 3.5, 8.2; sep. 5.8″) through a high-power eyepiece, aim your telescope just one degree east-south-east of it, at a point halfway between δ and another binary, 63 Gem (R.A. 07h 27m, Dec. +21° 27′). Nothing there? Not much, but it's a historic sky location. In January 1930, Pluto's faint image occupied this spot on a photographic plate taken at Flagstaff Observatory by Clyde Tombaugh. Since that time, Pluto has completed little more than one-third of its unconventional 249-year orbit around the Sun, and at the time of writing (January 2001) it lies in eastern Ophiuchus, more than 10° above the ecliptic.

Hercules (May–June)

51	M92 (NGC 6341)	RA 17h 17m	Dec +43° 08′

This fine globular cluster, about 30 000 l.y. away, is very much a "neglected neighbour", outclassed by M13, nine degrees to the SW. M92 is easy to find at mag. 6.1. Discovered not by Messier but by Bode in 1777, it is well worth a look, through any telescope. Do you see M92 as slightly oval?

52	NGC 6210 (Σ5)	RA 16h 45m	Dec +23° 49′

This mag. 9 planetary nebula, about 5000 l.y. away, is surprisingly bright. It looks like a slightly defocused star at high powers. Burnham saw it as bluish, and the Earl of Rosse listed it as "intense blue". I can't agree with these worthies as to the colour, but it's a rewarding target for small telescopes. The line from γ Her through β Her will take you to 6210, about 4° north-west of β. There is a double and a triple star in the low-power field of 6210: Σ2087 (mags. 8, 8, sep. 6") to the west, and Σ2094 (mags. 7, 8, 11, seps. 1.1" and 25") to the south-south-west.

Lacerta (September)

This small, unsung constellation deserves a better press. Its position within the Milky Way guarantees many intriguing clusters and fine starfields.

53	NGC 7209	RA 22h 05m	Dec +46° 30′

On first seeing this delicate cluster at ×108, I thought its ragged curves and clumps of stars suggested the outline of a maple leaf (Fig. 2.33). Distance about 3000 l.y.

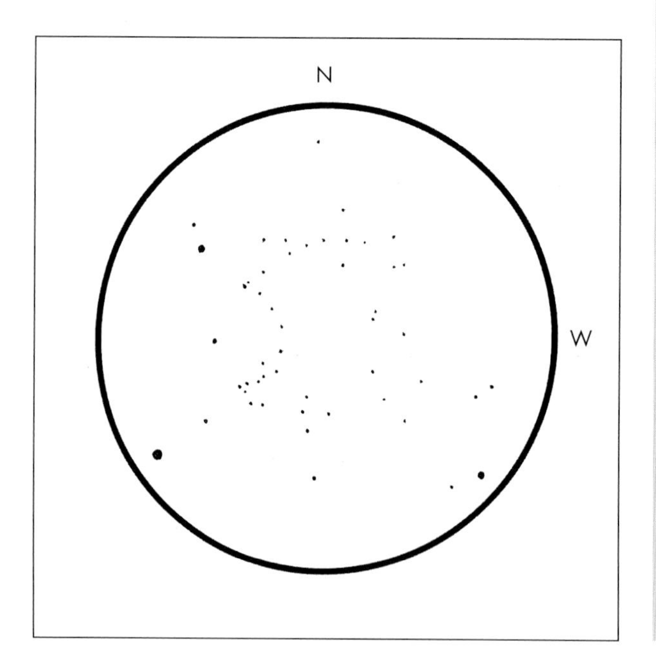

Fig. 2.33. NGC 7209.

Object: NGC 7209
Constellation: Lacerta
Type: Open Cluster
Magnitude: 7
Number of stars: ~50
***Norton's 2000.0* chart:** 13
***Uranometria 2000.0* chart:** 87
Instrument: 21-cm./8.5-in. reflector f/6
Magnification: ×108
Field diameter: 21′
Seeing: Ant. III
Light pollution: moderate
Date: 1994 September 1
Time: 2244 UT

Notes: Quite large, ragged cluster. Curves and 'points' suggest maple leaf.

54	NGC 7296	RA 22h 28m	Dec +52° 15′

A much tighter cluster than 7209, and closely following β Lac, which looks orange. The cluster has one dominant star (mag. ~9) which seems to be "sowing" several fainter members behind it as it moves across the field. Webb described the starfields of this area, with their chains, curves and pairs of faint stars, as "glorious", and they certainly repay sweeping with low and medium powers.

Leo (February–March)

55 56	3 Leonis 6 Leonis	RA 09h 28m RA 09h 32m	Dec +08° 10′ Dec +09° 45′

Two very similar, fairly wide double stars less than 2° apart. Intriguingly, 3 and 6 echo each other in orientation (p.a. 80°, 75°), magnitudes (6, 10; 6, 9) and colours (both yellow-orange and dull blue-grey). Separations 25", 37". Between them lies another double, ω (2) Leonis, much harder to split (mags. 6, 7; sep. 0.4"). High powers needed.

57	NGC 2903	RA 09h 32m	Dec +21° 30′

Leo contains some galaxies bright enough to show through moderate light pollution, and M65, M66 and M96 (all mag. 9) and M95 (mag. 10) are well known to many observers. Leo has many other galaxies attainable with medium to high powers in very clear skies: just in front of the Lion's "nose", search for the Sc system NGC 2903, a mag. 9.7 smudge with a comparatively bright nucleus (Fig. 2.34).

Leo Minor (February–March)

A small, neglected constellation with some interesting double stars.

58	Σ1374	RA 09h 41m	Dec +38° 56′

A binary with fine colours (mag. 7, yellow, and mag. 8, blue), though close (3.2"). Well seen at ×323.

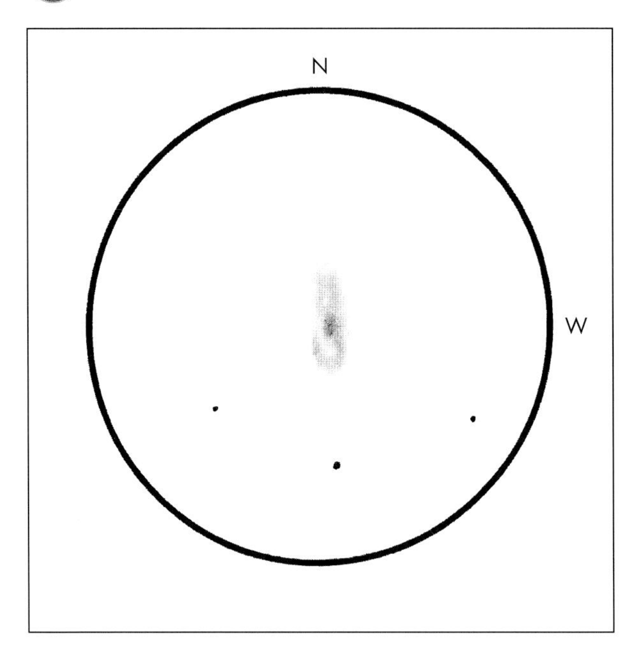

Fig. 2.34. NGC 2903.

Object: NGC 2903
Constellation: Leo
Type: Galaxy Sc
Magnitude: 9
***Norton's 2000.0* chart:** 7
***Uranometria 2000.0* chart:** 143
Instrument: 21-cm./8.5-in. reflector f/6
Magnification: ×108
Field diameter: 21′
Seeing: Ant. II
Light pollution: moderate
Date: 1992 March 9
Time: 0018 UT

Notes: Bright centre and neat oval form, with hint of arm to south.

59	NGC 3245	RA 10h 27m	Dec +28° 33′

Probably Leo Minor's best galaxy, mag. 11, rather elongated with a starlike nucleus at ×108. Why did D'Arrest call it "oblong"? One way to "fish up" this galaxy is to centre γ Leonis, in the "mane" of the larger Lion, in a telescope field, then move nine degrees northwards and two degrees eastwards.

Lynx (January–February)

Devoid of bright stars, Lynx has one or two surprises awaiting the dogged observer.

60	NGC 2419	RA 07h 38m	Dec +38° 53′

You would think that this globular cluster, so far away (~250 000 l.y.) that it is considered an "extragalactic wanderer", would be an impossible object, yet I saw it with averted vision from my suburban garden at ×108 on a very clear, still night even before those "sky-friendly" lights appeared in the street. There is a very slight hint of central brightening. Burnham gives its magnitude as 11.5, but it is easier to see than NGC 7006, another distant globular of the same estimated magnitude in Delphinus. Two mag. 8 stars, one of them double, point to its position closely following them in R.A. These stars are easy to find, as they are exactly seven degrees north of Castor and a little east.

Fig. 2.35. NGC 2683.

Object: NGC 2683
Constellation: Lynx
Type: Galaxy Sb
Magnitude: 10
***Norton's 2000.0* chart:** 7
***Uranometria 2000.0* chart:** 102
Instrument: 21-cm./8.5-in. reflector f/6
Magnification: ×108
Seeing: Ant. I
Light pollution: moderate
Date: 1995 December 26
Time: 0018 UT

Notes: A broad spindle, no bright nucleus, faint star at northern end. Temperature at observatory –8° C, air very still.

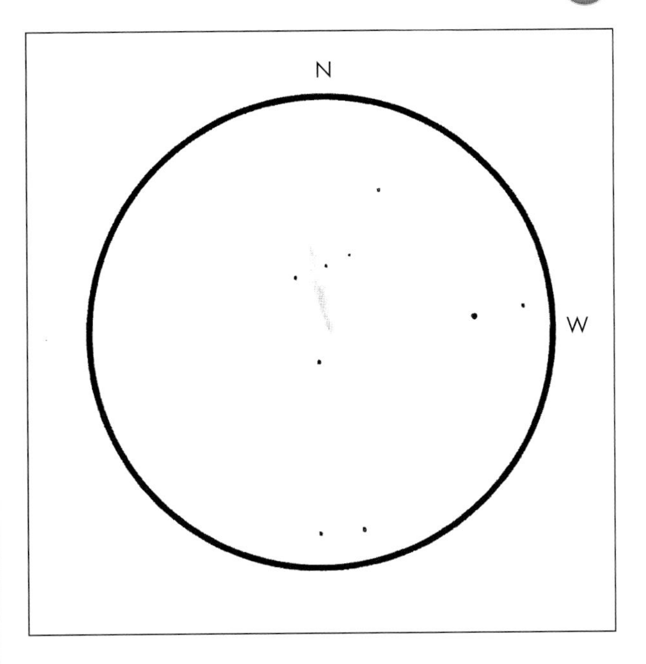

61	NGC 2683	RA 08h 54m	Dec +33° 30′

On a freezing night in December 1995, with the air temperature down to –8° C (17° F) and the sky remarkably clear, I found this broad spindle of a galaxy. It showed some detail at ×108, its southern end appearing slightly brighter than its northern end (Fig. 2.35). At mag. 10, its apparent size is 9′ by 1.3′ in a large telescope (Burnham). It lies about 20 million l.y. away.

62	19 Lyncis (Σ1062)	RA 07h 23m	Dec +55° 17′

Fairly bright and wide (mags. 5.5, 6.5, sep. 15"), this pair has an 11th magnitude companion at 74", p.a. 287° (i.e. preceding).

Lyra (June–July)

63	α Lyrae (Vega)	RA 18h 37m	Dec 38° 78′

According to Burnham, the first star to be photographed, from Harvard in 1850. The bluish optical companion (mag. 10, distance just over 1′) is almost lost in the glare of Vega (mag. 0.0). Glimpse it at highest powers through a 21-cm/8-inch instrument, though darker skies seem paradoxically to conceal it.

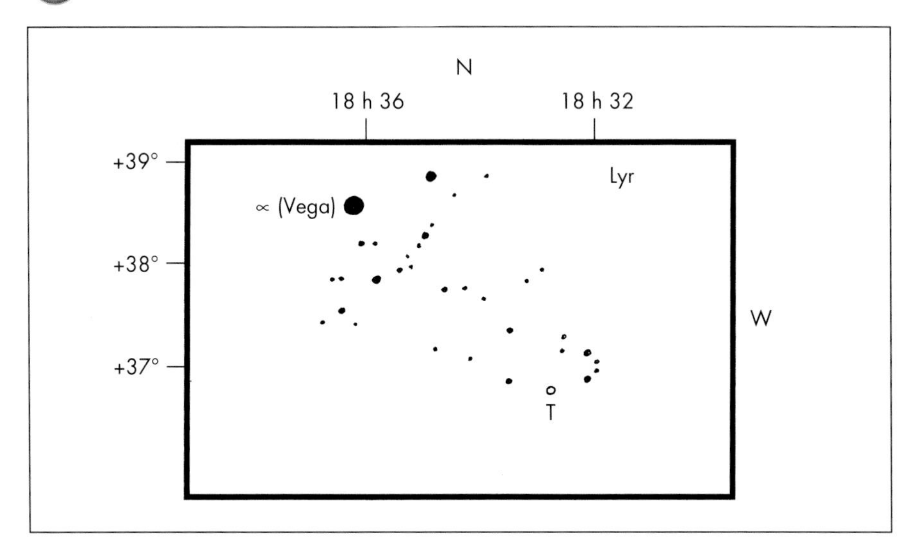

64	T Lyrae	RA 18h 32m	Dec +37° 00′

This blood-red jewel, when at its brightest (mag. 7.5 to 9.3, irregular), is a stunning object in a well-populated field. Do you agree that a fairly low power (say, ×50) seems to show off the colour best? Try sweeping slowly for T Lyr in the area 2° south-west of α (Vega), and let its colour signal its identity, or "star-hop" to the variable using the chart (Fig. 2.36), which shows stars (to mag. 9) in the field between Vega and T Lyr.

Fig. 2.36. From Vega to T Lyr.

Object: T Lyrae
Type: Irregular variable
***Norton's 2000.0* chart:** not shown
***Uranometria 2000.0* chart:** 117

Ophiuchus (May–June)

65	IC 4665	RA 17h 46m	Dec +05° 43′

A fine cluster for binoculars or a wide-field eyepiece, IC 4665 is wider than the full Moon, and easily found by panning 1.5° north-east from nearby β Oph. Alan MacRobert pointed out (*Sky & Telescope*, June 1989, p. 605), that the pattern of the main stars forms the word "HI" when south-west is up (Fig. 2.37). Perhaps a good object to "break the ice" when showing your floodlight-dependent neighbour, at the telescope, why you'd like the sky to be darker? IC 4665 is about 1000 l.y. away.

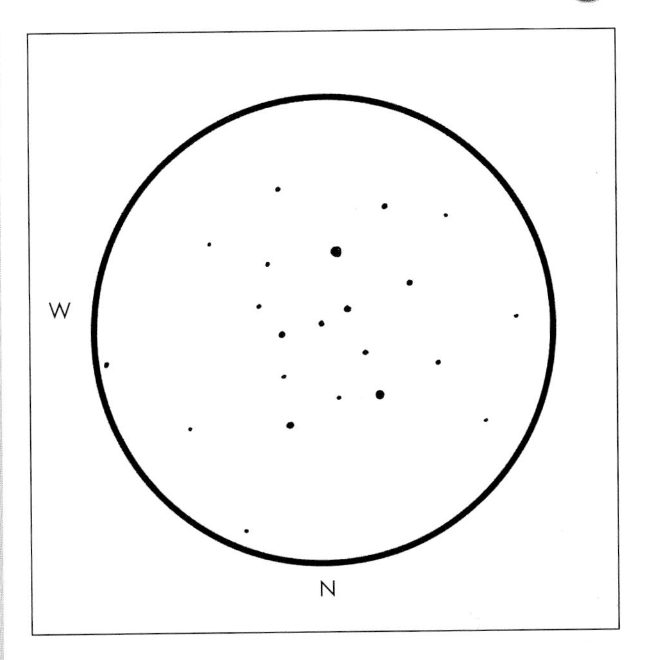

Fig. 2.37. IC 4665, the "HI!" cluster.

Object: IC 4665
Constellation: Ophiuchus
Type: Open Cluster
Magnitude: 4.5
Number of stars: 20
***Norton's 2000.0* chart:** 11
***Uranometria 2000.0* chart:** 203
Instrument: 21-cm./8.5-in. reflector f/6
Magnification: ×50
Field diameter: 45′
Seeing: Ant. II
Light pollution: moderate
Date: 1996 July 23
Time: 2300 UT

Notes: A grid pattern of bright stars in a very large cluster.

66	NGC 6633	RA 18h 28m	Dec +06° 34′

This compact curve of stars, of mags. 8 and below, became briefly the astrophotographers' favourite target on 1987 November 14, when the nucleus of Comet 1987s (Bradfield) passed right in front of it. The chart (Fig. 2.38) records the event as seen in a slightly hazy sky, at 18.48 UT, through a small 6-cm/2.5-inch refractor at ×30. Even without such distinguished company, NGC 6633 is still worth a look, through big (×20) binoculars or a low-power eyepiece. About 1000 l.y. away.

Orion (December–January)

67	σ Orionis	RA 05h 39m	Dec –02° 36′

A splendid multiple system. Easy to find, one degree south-west of ζ Ori, the left-hand "belt" star. You can see five components (mags. 4, 6, 6.5, 7.5, 10) at ×108, with the triple Σ761/761b close by to the north-east. σ Ori is 1400 l.y. away.

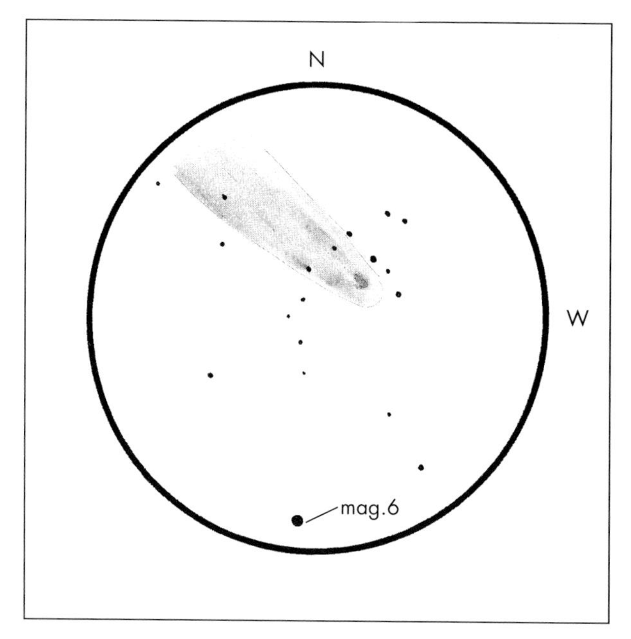

Fig. 2.38. NGC 6633 and Comet 1987S (Bradfield).

Object: NGC 6633
Constellation: Ophiuchus
Type: Open Cluster
Magnitude: 5
Number of stars: ~65
***Norton's 2000.0* chart:** 13
***Uranometria 2000.0* chart:** 205
Instrument: 6-cm./2.5-in. refractor
Magnification: ×30
Field diameter: 45′
Seeing: Ant. III
Light pollution: moderate
Date: 1987 November 14
Time: 1848 UT

Notes: A superb sight. Estimated magnitude of comet 5.5.

68	14 (OΣ98) Orionis	RA 05h 08m	Dec +08° 30′

This close binary, sep. 0.8″, looks unremarkable: with highest power, two fairly similar white stars (mags. 6, 6.5) can be seen, with a delicate little double (Σ643) close by to the south. But sketch 14 Ori over the years, and watch the position angle change. Its 160-year period means that, in the decade 2001–2010, the secondary will appear to move through more than 20° in its orbit around the primary. More information, and chart, in Karkoschka (see bibliography).

Pegasus (September–October)

69	NGC 7331	RA 22h 37m	Dec +34° 25′

At mag. 9.7, the brightest member of a whole field of galaxies to the north of η Peg, this almost edge-on Sc spiral (some authorities classify it as Sb) appears slightly asymmetrical in two ways: the obvious nucleus a little off centre, and the "spindle" appearing brighter along its southern "spike" (at ×108). Do you agree? Astronomers often propose 7331 as an example of what our Milky Way galaxy might look like from a viewpoint in deep space. Don't expect to see its very faint neighbours with modest instruments! Distance about 55 million l.y. (Fig. 2.39).

Fig. 2.39. NGC 7331.

Object: NGC 7331
Constellation: Pegasus
Type: Galaxy Sc
Magnitude: 10
***Norton's 2000.0* chart:** 3
***Uranometria 2000.0* chart:** 123
Instrument: 21-cm./8.5-in. reflector f/6
Magnification: ×108
Field diameter: 21′
Seeing: Ant. II–III
Light pollution: moderate
Date: 1988 November 12
Time: 2155 UT

Notes: Slightly asymmetrical, more visible to south. Nearby fainter galaxies not seen.

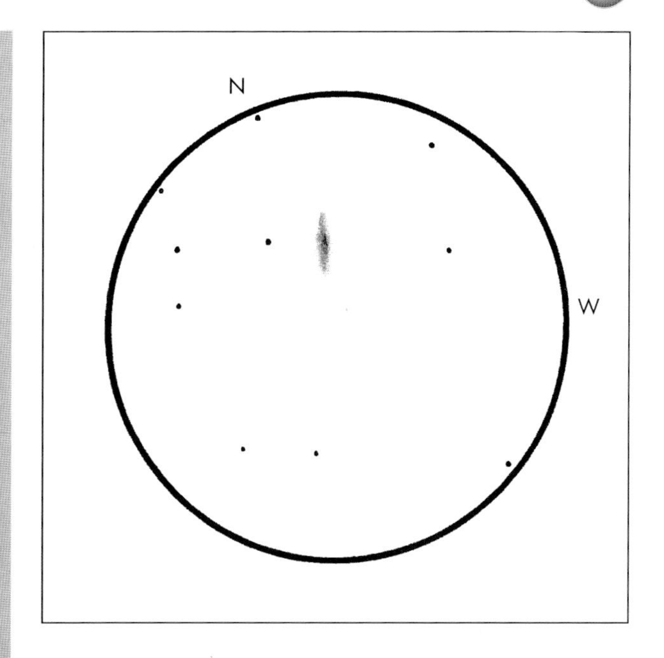

70	3 Pegasi	RA 21h 38m	Dec +06° 37′

A glittering field, three degrees south and nearly two degrees west of ε Peg: 3 Peg (mags. 6, 8.5, sep. 39", and there is a third, mag. 13 member) is attended to the north-east by fainter OΣ443 (8, 8.5, 8"). Other nearby stars combine to give the impression of a sparse cluster. Medium power.

Perseus (November–December)

Perseus' crowds of stars offer countless targets, both famous and obscure, even in poor skies. Have you ever come upon these three?

71	Σ297	RA 02h 46m	Dec +56° 30′

A pretty triple star, in a crowded field, colours (blue-white, blue-white, yellow) well seen at ×108. Mags. 8, 8.5, 10.5. One degree away, south preceding, is Trumpler 2, a delicate cluster. To find this triple, just savour the delights of the Perseus Double Cluster (NGC 869/884), then move half a degree to the south, and Σ297 follows closely in R.A.

72	NGC 1582	RA 04h 33m	Dec +43° 50′

It's easy to find this large cluster, nearly zenithal at Christmastime, seven degrees preceding the giant star ε Aurigae. Its brightest members form a straggly "S", snaking across the 3/4° field of my ×50 eyepiece. There is some rewarding sweeping, with many pairs and curious star alignments, both north and south of the cluster.

73	Stock 4	RA 01h 53m	Dec +57°

St 4 is a grand sight at ×50. Not listed in most handbooks and atlases, this large, loose group lies neglected in the north-east corner of Perseus. Half the 45′ low-power field (Fig. 2.40) is strewn with faint stars, and my first impression of it through the telescope on a clear, frosty autumn night, was: "like a rain of tiny ice crystals". Easily found, as its declination is the same as that of the Double Cluster, which it precedes by seven degrees (28 min in R.A.).

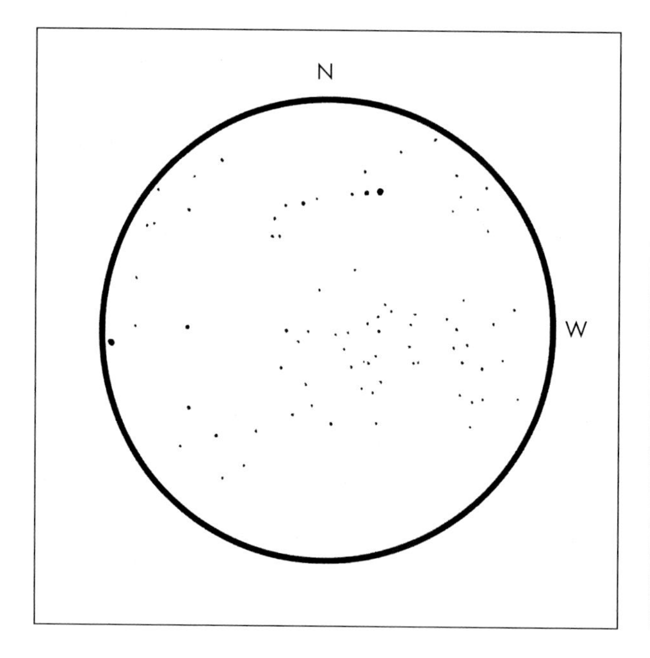

Fig. 2.40. St 4.

Object: Stock 4
Constellation: Perseus
Type: Open Cluster
Magnitude: 7
Number of stars: 100+
***Norton's 2000.0* chart:** not shown
***Uranometria 2000.0* chart:** 37
Instrument: 21-cm./8.5-in. reflector f/6
Magnification: ×50, ×108
Field diameter: 45′
Seeing: Ant. II
Light pollution: moderate
Date: 1991 December 4
Time: 2234 UT
Notes: A mass of faint stars.

Pisces (September–October)

74	ψ^1 Piscium (Σ88)	RA 01h 06m	Dec +21° 30′

Superb white-white twins, both mag. 5, and 39" apart. A very easy object at whatever power.

Sagitta (July–August)

The little Arrow's position in the Milky Way guarantees plenty of interest packed into only 80 square degrees – only one northern constellation is smaller: Equuleus, the Foal, at 72 square degrees.

75	U Sagittae	RA 19h 18m	Dec +19° 37′

The eclipsing binary Algol (β Per) has a serious if shy rival here. Follow the variations of U Sge, an easy binocular object, by comparing it with Σ2504 (mag. 6.5), half a degree to the south-east. The field is easy to find, just west of the "hook" of the well-known Coathanger asterism Collinder 399 in Vulpecula (Fig. 2.41). At maximum, U Sge (mag. 6.4) is a little brighter than Σ2504. Every 81.14 hours (3.381 days), it will dim rapidly towards a minimum of mag. 9.2, where it remains for 1h 40m. The brighter, blue B8-type component is completely hidden at each eclipse by its faint, Sun-like G2 companion; Algol's "eclipse" is only partial, with about 80% of the brighter star hidden by its large neighbour.

76	Field of 13 Sagittae	RA 20h 00m	Dec +17° 30′ (mags. 6, 12; sep. 28″);

77	Field of 15 Sagittae	RA 20h 04m	Dec +17° 05′ (mag. 6);

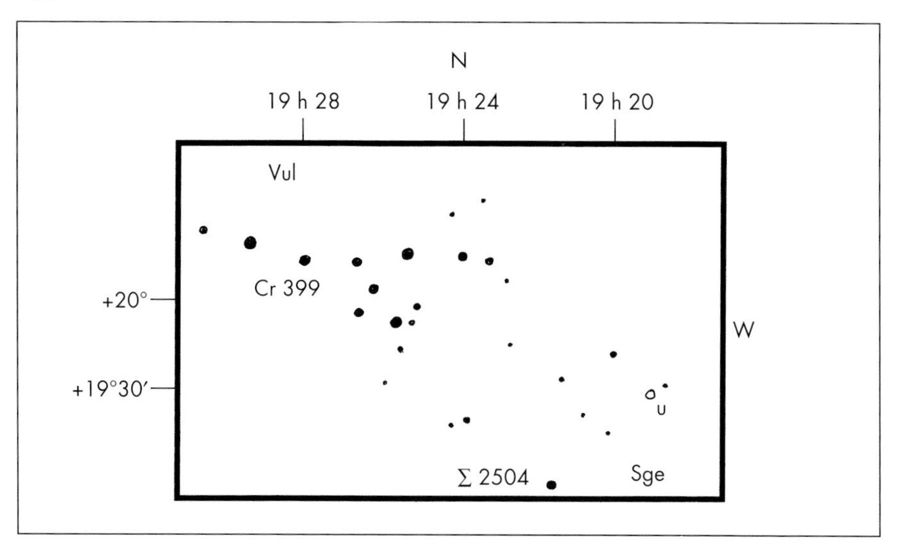

Fig. 2.41. Finder chart for U Sge.

Object: U Sagittae
Type: Eclipsing binary
Period: 3.381 days
***Norton's 2000.0* chart:** 13
***Uranometria 2000.0* chart:** 161

78	WZ Sagittae	RA 20h 07m	Dec +17° 40′.

After exploring the two striking little crowds of stars around 13 and 15 Sge, one degree apart in a rich region of the Milky Way and easy to find just two degrees below γ Sge, the "point" of the Arrow, examine the position of another recurrent nova, WZ Sge, two degrees west of 13 Sge. We may be due for another outburst of WZ, which is normally invisible at mag. 14 to 16: it brightened to mag. 7.0 in 1913, to 7.7 in 1946, and to 8.6 in 1978. According to Burnham, the light curve and spectrum of WZ Sge suggest that it is a binary consisting of a red dwarf and a white dwarf in a very tight orbital dance: they may be closer to each other than the Earth is to the Moon. The chart (Fig. 2.42) shows stars for "hopping" from γ Sge through 13 to WZ.

Scutum (June–July)

Although the Shield may never climb very high in your sky, and you may not see the wonderful Milky Way condensation which is its starcloud from a light-polluted site, try:

79	Σ2391	RA 18h 49m	Dec –06° 01′

This fine double star (mags. 6.5, 9.5, sep. 12″), just over one degree south of β Sct, is actually a triple, but there is no chance of glimpsing the mag. 14 companion with modest instruments. The brighter of the two visible stars looks unusually white. Moderate power.

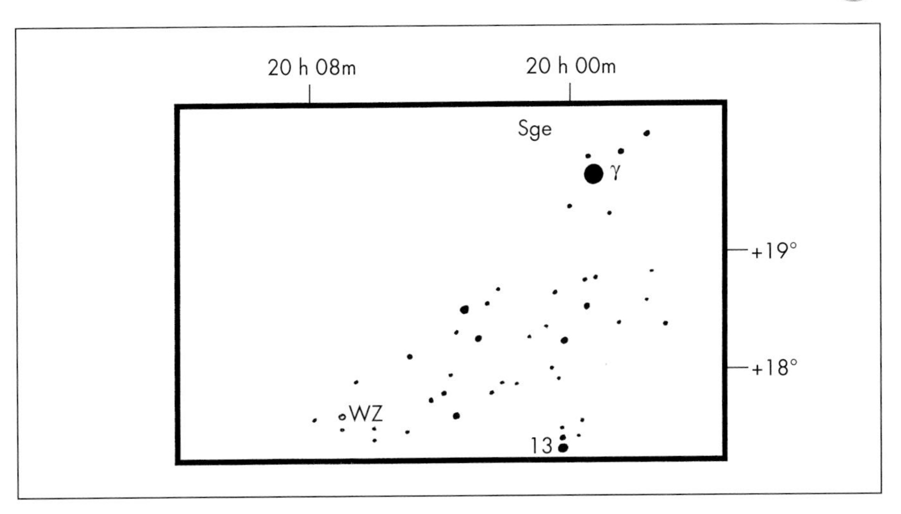

Fig. 2.42. Finder chart for WZ Sge.

Object: WZ Sagittae
Type: Recurrent nova
***Norton's 2000.0* chart:** 13
***Uranometria 2000.0* chart:** 163

Serpens (May–June)

The divided Snake offers many famous targets. Thanks to the Hubble Space Telescope, the "Pillars of Creation", the dust towers of the Eagle Nebula (M16), are now probably the world's best known deep-sky object. Serpens' lesser known pleasures include:

80	R Serpentis	RA 15h 51m	Dec +15° 08′

Easy to find with binoculars between β and γ Ser (Fig. 2.43) when at its maximum of mag. 6.9, R Ser is a typically red long-period variable. Its period is 356 days. At its minimum, it will have faded to about mag. 13. It is about 1000 l.y. away.

81	IC 4756	RA 18h 39m	Dec +05° 25′

This extensive cluster, 1400 l.y. away, appears twice as wide as the full Moon. It is worth looking at with both high powers and very low powers, or binoculars. High magnification brings out many stars seemingly crowding into its central region, enclosed in a rough pentagon of its brightest members, while a ×20 binocular field renders the surprisingly large cluster in its entirety. NGC 6633 in Ophiuchus (see above) is three degrees away to the north-west.

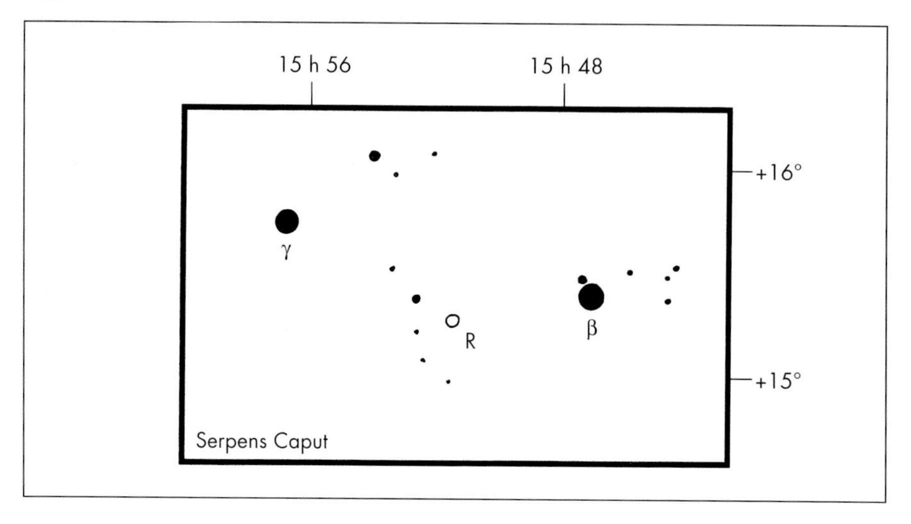

Fig. 2.43. Finder chart for R Ser.

Object: R Serpentis
Type: Long-period variable
Period: 356 days
***Norton's 2000.0* chart:** 11
***Uranometria 2000.0* chart:** 200

Taurus (November–December)

82	NGC 1647	RA 04h 48m	Dec. +19° 03′

A really "neglected neighbour", completely outclassed by the nearby Hyades, but still a fine object in its own right. With a low power, two main streams of stars are obvious, with a suggestion of many more just beyond the limit of vision. With south upwards, through an inverting telescope, the fanciful observer might see NGC 1647 as a pyramid standing on its apex. See chart for HU Tauri (below).

83	Σ670	RA 05h 17m	Dec +18° 27′

A high power will reveal the dual nature of this system (mags. 7.5, 8; sep. 2.5"), and possibly its colours (white, blue), but the intriguing thing about Σ670 is the "driveway" of several faint stars leading up to it from the south. Not far away northwards is:

84	Σ680	RA 05h 19m	Dec +20° 10′

...with a dull red mag. 6 primary and a difficult mag. 10 secondary (sep. 9"), at the start of a pretty little east–west chain of stars containing: *(see next item)*

85	Σ674 (CD Tau)	RA 05h 18m	Dec +20° 10′: mags. 6,9; sep. 10″.

There are several pairs in this region: try sweeping with a medium power.

86	HU Tauri	RA 04h 38m	Dec +20° 40′

A relatively bright eclipsing binary, normally at mag. 6.0 and easy to locate: four degrees almost due north of Aldebaran, and exactly halfway between ε and τ Tau. At intervals of 49.4 hours (2.06 days), it dims to mag. 6.8, the whole event being well within the range of small binoculars. The chart (Fig. 2.44) shows stars between Aldebaran and HU Tau, with NGC 1647 (see above) in the field.

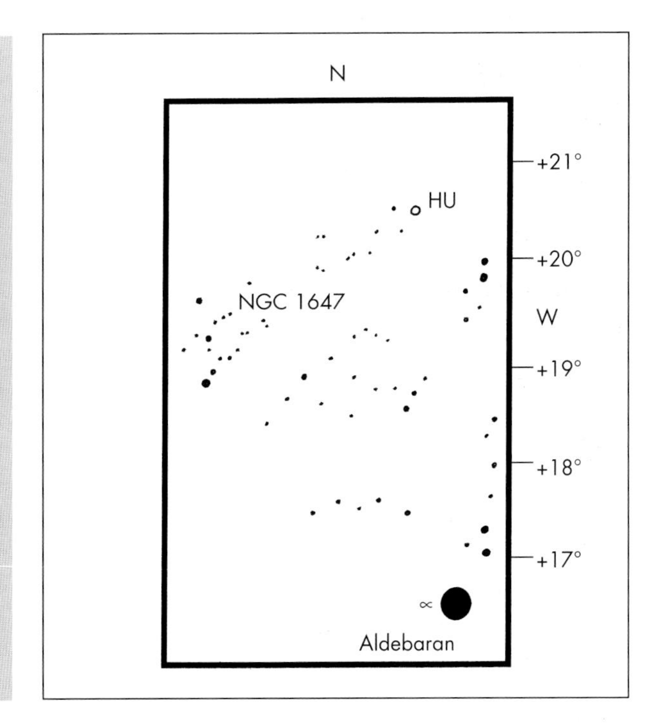

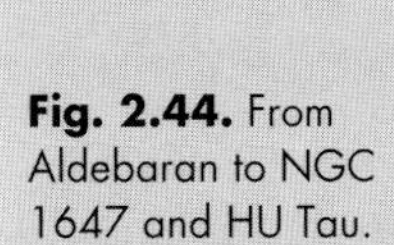

Fig. 2.44. From Aldebaran to NGC 1647 and HU Tau.

Object: HU Tauri
Type: Eclipsing binary
Period: 2.06 days
***Norton's 2000.0* chart:** 5
***Uranometria 2000.0* chart:** 134

Triangulum (October–November)

87	M33 (NGC 598)	RA 01h 34m	Dec +30° 39′

The main attraction of Triangulum, the extensive face-on spiral galaxy M33, is a notoriously difficult object to locate with moderate instruments because of its low surface brightness. Even in good dark skies, a thin haze or slight moonlight can veil it, and observers who suffer from any degree of light pollution might decide not to bother. Also, its large size (90′ by 60′) can throw some observers who are expecting to find something smaller. Imagine my surprise then, when, on the night of 1989 October 5, I chanced a look at it in my moderately light-polluted sky through the 21-cm/8-inch and saw some ghostly but real detail! *Exceptional* sky clarity seems to be the key here so, on rare, clear autumn nights, try M33. I used ×50, and sketched a vaguely "spiral" core and a few nebulous knots (Fig. 2.45). A memorable evening. A borrowed broad-band filter improved the view slightly.

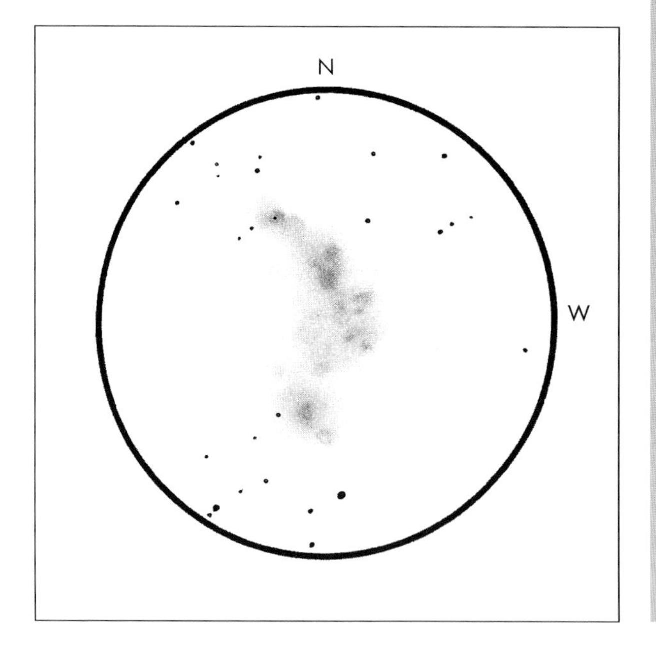

Fig. 2.45. M33.

Object: M33 (NGC 598)
Constellation: Triangulum
Type: Galaxy Sc
Magnitude: 6.7
***Norton's 2000.0* chart:** 3
***Uranometria 2000.0* chart:** 91
Instrument: 21-cm./8.5-in. reflector f/6
Magnification: ×50
Field diameter: 45′
Seeing: Ant. I–II
Light pollution: moderate
Date: 1989 October 5
Time: 2247 UT
Notes: Large, blotchy object, with a few 'highlights'. Slight indication of spirality. Filter.

88	ι (6) Trianguli	RA 02h 12m	Dec +30° 18′

Here is a fairly bright binary (mags. 5.3, 6.9) with a strong colour contrast. It is easy to split with a high power (sep. 4″). Victorian astronomer Admiral William Smyth called the colours "exquisite", and chronicler of the stars Robert Burnham Jr. saw them as a definite yellow and blue pair. Spectroscopy splits them both again. ι Tri is 300 l.y. away.

89	Σ239	RA 02h 17m	Dec +28° 46′

From ι Tri, move one and a half degrees southwards, and then one degree east, to find this binary. The brighter component is listed as mag. 7, and its companion mag. 8, though they look fairly similar in brightness to me. What distinguishes them is their colours, with the primary a fairly unsurprising yellow, but the other star, according to my notebook, an "elusive steely-grey" (?) colour at ×108. Nineteenth-century observers were also unsure: Reverend Webb saw the secondary as "bluish-grey", and Franks as "lilac".

Ursa Major (February–March)

In early spring, the Great Bear prowls high above late at night in mid-northern latitudes. The most extensive of the exclusively northern constellations, covering 1280 square degrees, it is a fertile area for seeking lesser-known deep-sky objects.

90	σ^2 Ursae Majoris	RA 09h 10m	Dec +67° 10′

This is another binary whose orbital changes you can follow, with high magnification, as the years go by. Gold and greenish-white, the two components are close, sep. 2.4" (increasing). The 1917 edition of Webb's *Celestial Objects for Common Telescopes* gives the position angle for 1912 as 147°, M. Duruy measured it at 45° in 1943, and *Burnham's Celestial Handbook* has the 1962 p.a. as 20°. The secondary has now passed its northerly position (p.a. 0°) as it pursues its 700-year swing around the primary.

91	23 Ursae Majoris	RA 09h 31m	Dec +63° 04′

23 UMa is an unremarkable double, mags. 3.7, 8.9, sep. 22". The challenge lies in spotting an optical neighbour, a mag. 10 star at 99″ south preceding. In my ×50 field (diameter of field 45′), a line from the primary through this faint companion points towards the compact elliptical galaxy NGC 2880 (see below).

92	NGC 2880	RA 09h 30m	Dec. +62° 30′

For a galaxy of mag. 12.7 (Burnham), 2880 was surprisingly clear on a calm night through the medium-power ×50 eyepiece (Fig. 2.46). It looked strongly pear-shaped, widening towards the preceding end, and of uniform brightness right across. The line from 23 UMa, if continued through the galaxy, leads to a pleasing white triple star, 4 arc minutes away. I have been unable to discover its designation.

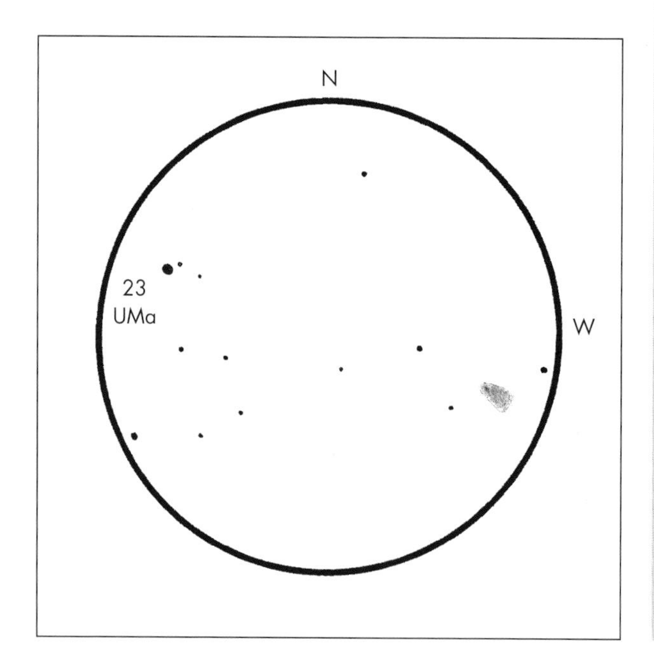

Fig. 2.46. From 23 UMa to NGC 2880.

Object: NGC 2880
Constellation: Ursa Major
Type: Galaxy E3
Magnitude: 12
***Norton's 2000.0* chart:** 1
***Uranometria 2000.0* chart:** 23
Instrument: 21-cm./8.5-in. reflector f/6
Magnification: ×50
Field diameter: 45′
Seeing: Ant. I–II
Light pollution: moderate
Date: 1990 July 25
Time: 2318 UT

Notes: Faint and featureless, pear shaped. At limit of vision.

93 94	83 Ursae Majoris Σ1795	RA 13h 41m RA 13h 59m	Dec +54° 40′ Dec +53° 05′

These two objects form part of the string of stars of mags. 5–7 curving away from Mizar (ζ UMa) towards ι and κ Boötis. 83 UMa is a late-stage red star, mag. 4.7, spectrum M2, and usually fairly unremarkable. Burnham recommends an occasional look at the area while observing, since, on 1868 August 6, Birmingham recorded a short-lived outburst by this star. Its apparent brightness had increased to that of δ UMa, a rise of 1.4 magnitudes to mag. 3.3. By August 7, it had sunk back to its previous state.
Σ1795 (mags. 7, 9.5; sep. 7.7") has a fine white primary, and the faint secondary can appear bluish. A neat pair at ×108.

95	β918	RA 11h 58m	Dec +32° 20′

A run-of-the-mill binary, with one star much brighter than the other, in a sparsely populated field? β918 (mags. 7, 13; sep. 7.5") looks just that, but on a *really* clear night, without moonlight or haze and with good seeing, use averted vision to seek three faint smudges preceding in R.A., all galaxies: NGC 3991, 3994 and 3995 (Arp 313), in the same field even with a fairly high power.

96	NGC 3992	RA 11h 57m	Dec +53° 22′

You might assume that 3992, a mag. 10.9 barred spiral galaxy, is part of the same group as 3991, 3994 and 3995 (above), but the great extent of Ursa Major becomes apparent when you realise that it is 21° to the north of them, in the same low-power field as γ UMa (Phad). Sometimes referred to by Owen Gingerich's 1960 appellation M109, it is a beautiful object in photos, shaped like a slewed Greek θ. There is a fine photo of NGC 3992 in *Sky & Telescope*, July 1985, p. 32. The sketch (Fig. 2.47) shows my impression of it on a clear February night in 1992, when it was almost at the zenith. I described the large inner brightness, which bulges slightly northwards, as "bean-shaped"; a fainter oval glow surrounds this at ×108. Distance 60 million l.y.

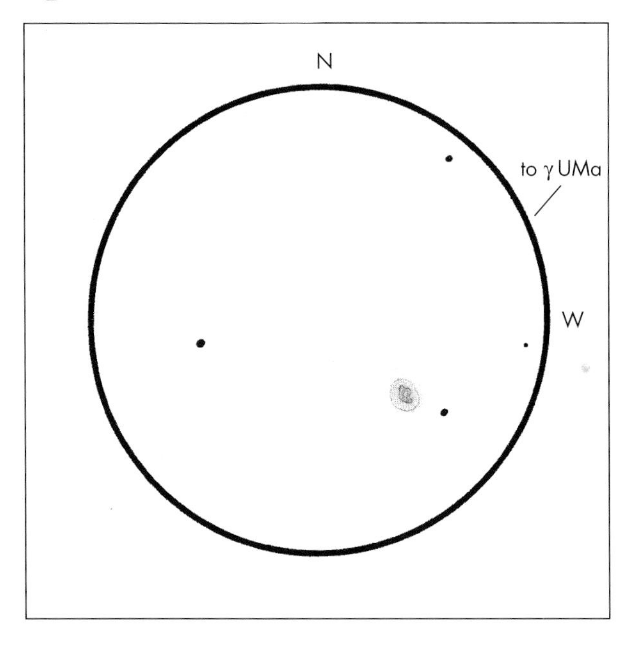

Fig. 2.47. NGC 3992.

Object: NGC 3992
Constellation: Ursa Major
Type: Galaxy SBb
Magnitude: 10.9
***Norton's 2000.0* chart:** 9
***Uranometria 2000.0* chart:** 47
Instrument: 21-cm./8.5-in. reflector f/6
Magnification: ×108
Field diameter: 21′
Seeing: Ant. III
Light pollution: moderate
Date: 1992 February 4
Time: 2320 UT
Notes: Some structure apparent, bean shaped nucleus, bulging to SW.

Ursa Minor (May–June)

97	π^1 Ursae Minoris (Σ1972)	RA 15h 29m	Dec +80° 28′

A very easy binary for a low-power eyepiece (mags. 6.5, 8; sep. 31"), with a faint (optical?) companion (mag. 11.5) at p.a. 104°. Both main stars are sunlike, both yellowish, though Franks saw the mag. 8 as "bluish-white". Nearby π^2 (Σ1989) is another binary, but it is difficult to split at 0.7" (1975 estimate). Look 2.5° north from ζ (mag. 4.3), and 15 minutes forward in R.A.

Virgo (March–May)

Virgo holds more than just galaxies! In my trawls through this constellation, steeped in skyglow from Poole, over the years, in a usually fruitless effort to find many of the hundreds of galaxies shown on charts such as *Uranometria 2000.0*, I've encountered other things:

98	17 Virginis	R.A. 12h 23m	Dec. +05° 20′

Failing to fish up any of the dim galaxies which crowd the field, I found some consolation in this binary (mags. 6.5, 9; sep. 19″). The primary is a rather dull white and, although there are some intriguing reports in Webb's *Celestial Objects* of colours as various as orange, blue and purple for the secondary, it looks much the same as its neighbour to me. Moderate power.

99	γ Virginis (Porrima)	RA 12h 42m	Dec –01° 26′

A showpiece among the visual binaries, with twin yellow F-type stars (both mags. 3.65). γ Vir, which Burnham compared in a striking analogy to the distant headlights of a car approaching you from space, has been closing fast since its maximum 6" separation in 1920, and will appear single through amateur telescopes as 2007 approaches, when the stars will be a mere 0.3" apart. When will you first be able to split this pair as they move away towards apastron? Erich Karkoschka's information-packed *Observer's Sky Atlas* lists two-yearly separations for this binary until the year 2015, by which time it will have opened to 2.2".

Vulpecula (July–August)

100	NGC 6940	RA 20h 35m	Dec +28° 15′

The Veil Nebula (NGC 6960 and 6992), the brightest part of which is at the absolute limit of vision from my observing site, shares its niche below the eastern wing of Cygnus the Swan with this large, attractive cluster of many faint stars just to the south-west of it and across the border in Vulpecula. NGC 6940 can be surprisingly bright through 10 × 50 binoculars. Through the telescope, with a fairly low power, you first see a line of three brighter stars, and a neat pair 20′ away; but use averted vision, and the faint riches between come to light (Fig. 2.48). The area within the dashed line on the sketch was a mass of unresolved "stardust" at ×50. The cluster is 1500 l.y. away.

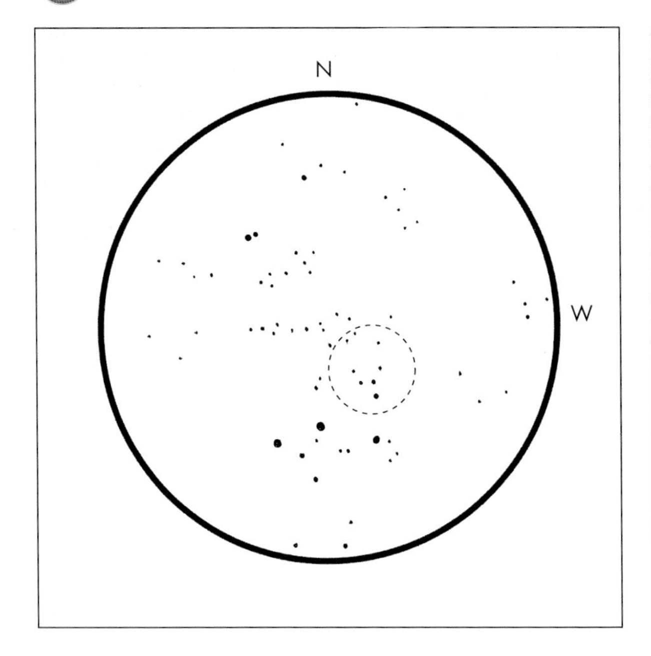

Fig. 2.48. NGC 6940.

Object: NGC 6940
Constellation: Vulpecula
Type: Open Cluster
Magnitude: 6.5
Number of stars: ~100
***Norton's 2000.0* chart:** 13
***Uranometria 2000.0* chart:** 120
Instrument: 21-cm./8.5-in. reflector f/6
Magnification: ×50
Field diameter: 45′
Seeing: Ant. I
Light pollution: moderate
Date: 1997 July 25
Time: 2334 UT

Notes: Very rich aggregation in southern half of cluster. A sparkling field.

If you decide to try for some (or all!) of the objects in this list, I hope that finding them and forming your own impressions will give you as much pleasure as I have had in choosing them from my frayed notebooks, reminiscing on nights well spent beneath the stars.

For updated information in the list above, I am indebted to Dr Andrew Hollis (British Astronomical Association Remote Planets Section), Guy Hurst (UK Nova/Supernova Patrol), Brian McInnerny (BAA Variable Star Section), Roger Pickard (BAA Variable Star Section), and Colin M. Pither (double stars). I would be grateful for any additional or corrected information on any of the objects at <bob@mizar-astro.freeserve.co.uk>

Chapter 3

Dark Future?

Fig. 3.1. A hill too steep to climb? Bath University's sports floodlights dominate the general town skyglow from high ground above the southern approach road (photo: Mike Tabb).

The task facing the astronomical community, of confronting and reversing the light pollution which has erased much of the detail of the night sky since the 1950s, seems to stretch uncertainly into the future, with many of those affected feeling that a small minority can never make themselves heard against mighty vested interests, erroneous public assumptions and even a perceived resignation among many astronomers to the fact that the hill is just too steep to climb (Fig. 3.1).

Fig. 3.2. Bob presents the Campaign for Dark Skies' Award of Appreciation to broadcaster John Humphrys, who has often involved himself in the dark skies debate (courtesy CfDS).

But the tide *is* beginning to turn. A professional lighting consultant, bemoaning the fashion for rural pubs to floodlight their exteriors, wrote in early 2001 that the astronomers' campaign had become a serious factor in lighting design, and a lighting industry marketing manager said that it was one of the most telling factors in lighting design change in the last decade.The IDA, CfDS and numerous other organisations (see Appendix 1) have made light pollution a talking point, not just among astronomers, but in the media (Fig. 3.2), in legislative assemblies, in lighting professionals' journals, and in legal chambers. Little children and experienced politicians know the meaning of the term. It now appears in dictionaries as a separate entry. Organisations as various as countryside preservation campaigns, ornithologists and guideline bodies at the forefront of the lighting and engineering professions are playing an active rôle in spreading the word that light is not necessarily always right, whatever the quantity and in whatever direction it goes. The media are treating the subject seriously: for example, in October 2000, BBC Radio 4's *Costing the Earth* programme devoted itself to the subject of light pollution. The main message of the broadcast was that the problem affects not just astronomers, but the ordinary householder, troubled by local sports floodlights to the extent that moving house is being considered; not just airline pilots, flying above a confusing urban glow, but road safety officers worried about glare from roadside entertainment complexes; and not just environmentalists decrying the urbanisation of the countryside, but medical experts investigating the deleterious effects on health and development of sleepers being bathed in artificial light.

To throw the brake on a runaway vehicle, it is first necessary to be on board. Public opinion will turn against light pollution only if information is available and the public educated about the problem. To bring the optimum night sky to everybody, whether they live in downtown Los Angeles or an isolated cottage in the Highlands of Scotland, it is first necessary to make sure that they know the basics of the light pollution debate: they need to understand the extent of the problem, what has been lost to them, and its value. How is this education to take place? The public will consider obtrusive light as a potential and actionable nuisance, and feel minded to do something about it, or curb their own excesses, only if they know what we have lost of our environment and quality of life to light pollution. Those who make, sell, choose and install lights should automatically incorporate control of the direction of emissions, and avoidance of glare and overlighting, just as those professionals who deal with domestic electrical fittings automatically put safety first. The political will of legislators to protect the environment above, with real action instead of just consolatory statements, must be sparked by raised public awareness of the need to curb the misuse of light, for without public demand and a lively national debate, legislation is unlikely to be enacted. Astronomers are everywhere, and can be educators in many ways; teachers, who are often the first environmental champions children encounter, should know of the problem of excess light, and have the information at hand to be able to mention it alongside other forms of environmental harm.

What needs to happen, then, to turn the minds of all these different people (the public, the lighting professionals, legislators and educators, be they astronomers or teachers) towards the problem? What action is possible? Will there be a twenty-first century solution to a twentieth-century problem?

3.1 Light Pollution: Solutions for the Twenty-First Century

What Should Manufacturers Be Doing About Light Pollution?

Are those who design and create today's lamps taking note of the calls by CIE, CfDS, IDA, IESNA, ILE and other bodies for more sensitive lighting? All national and international lighting companies know that they can take measures either to protect the environment from waste light, or spoil it. IDA reports that US manufacturers are slowly absorbing the message, but its translation into good practice is slow. UK companies are perhaps more willing to retool and adapt, and "sky-friendly" lamps are certainly appearing in large numbers on Britain's roads, but the private lighting sector lags woefully behind.

The *Lighting Journal* is the respected official publication of the Institution of Lighting Engineers (ILE). Since the ILE is a keynote body in the international lighting community, reading its professional journal is a good starting point for anyone wanting to "take the temperature" of the industry.

The June/July 1995 edition of the *Lighting Journal* can be used to sum up the drift towards environmentally sensitive luminaires which began in the last decade of the twentieth century. The very first double-page spread is an advertisement by a lighting company whose products are distributed worldwide. Most of the ad consists of a photo of a starfield. The rest praises a local council for choosing flat-glass road lamps, and mentions the British Astronomical Association's Good Lighting Award, presented to the council for "protecting the night sky with its pollution-free road lighting scheme ... directed evenly onto the road below, and none invades the sky above". Remember, this journal is read by the informed professionals of the industry and their potential customers, and not by the general public. The company in question obviously considers that the climate within the lighting community is one which approves of sensitive lighting, and that potential customers will see environmentally "caring" lights as a good idea, worth spending their money on. Such advertisements were soon to become quite common, culminating in double-page spreads in the national press in the late 1990s, explaining the need to combat light pollution and promoting the sky-friendly products of an Italian company. Stars are now quite a common backdrop in the publicity of the lighting industry (Fig. 3.3).

Next comes the *Journal*'s leading article, *Social Factors Behind the Development of Outdoor Lighting*, by Michael Simpson, then President of the Institution of Lighting Engineers and chief lighting engineer with a leading international manufacturer. The article begins with a review of the history and benefits of outdoor lighting, but soon, Mr Simpson writes with less enthusiasm about the "amorphous yellow glow ... spreading across the countryside ... and never mind the spill or quality". In his opinion, a "metamorphism" had been

Fig. 3.3. An advertisement at the Total Lighting Exhibition 2000, using the stars as a feature.

occurring within the lighting industry during the early 1990s: "We were learning that outdoor lighting is more than just filling the space with light; learning that it is more than just a way of making our roads visible to motorists; learning that sensitivity in design is equally as important outdoors as it is indoors; and learning to take care of our environment."

The environmental/astronomical community is seen as important and worth listening to:

> The environmentalists are concerned about the impact the equipment has on the landscape whether by day or night. In addition we have the astronomers, who are concerned about the amount of artificial light which is scattered in the atmosphere ... The astronomical lobby has been particularly effective in persuading us that direct upward light must be reduced ... With the recent passing of Halley's Comet [1985–86], one commentator records the fact that his grandparents witnessed the previous passing with the naked eye; he had to make do with television pictures. It has been calculated that the effect of skyglow over London will reduce the visibility of stars by four stellar magnitudes. This means that the Pole Star completely disappears ... for millions of potential stargazers. Bearing in mind the importance of the stars for many early civilisations and travellers, and the place of astronomy in the National Curriculum, this is not a legacy we should perpetuate.

Beneath these words appears a composite satellite view of the Earth at night, showing the tracery of wasted light rising from the world's inhabited areas. Mr Simpson concludes this section with the observation that "the roadlighting community has responded well to environmental pressure".

It is indeed true that "sky-friendlier" road lights are coming on stream in large numbers (Figs. 3.4, 3.5a/b), though the 30-year lifetime of the average lamp means that the replacement of old-stock, less well directed lighting cannot be an overnight process.

Driven by competition and by environmentally aware voices within the industry, by the demands of highway engineers for more directional lighting, and by the astronomers' continued publicising of the need for improvement, the climate within the road-lighting industry is warming. Members of the British Astronomical Association's Campaign for Dark Skies visited the UK lighting industry's main showcase exhibition, *Total Lighting 2000*, in May 2000, and we spoke to many of the exhibitors; the great majority of

Fig. 3.4. "Sky-friendlier" FCO lights (black casings) replace old LPS types at a rural roundabout.

outdoor luminaires displayed around the great hall of the National Exhibition Centre had good light control, and road lights and sports lights vied with each other to put their emissions where they are needed. The exhibition's introductory brochure featured a starry sky and a downward-directed floodlight (Fig. 3.6), and with the exception of a few exhibitors who sacrifice light control for quaint or startling design, I was encouraged by what I saw and heard. If it's getting better, why hasn't skyglow begun to disappear? Why does glare remain a feature of the night-time landscape? Why are so many people complaining about *new* lights which cause light trespass?

Part of the answer lies in the fact that things are not going in such a positive direction in other enclaves of the industry. Domestic and electrical retailers still offer basic, over-powered security lights, stacked high and sold cheap. Some of the retailers claim that their products are all vetted for environmental awareness. Environmentally sensitive light control within the average and by now "traditional" 300 W or 500 W home security light is non-existent (Fig. 3.7), and I have never seen a set of instructions inside one of these offering advice on sensitive mounting. The usual angle at which these are set is with the front glass at nearly

a

b

Fig. 3.5 a Glare and skyglow from a rural roundabout lit by old LPS lamps **b** The same scene after refitting with cut-off lamps: sideways glare and skyglow are much reduced (photo: John Ball).

90° to the ground, and many such lamps have their passive infrared (PIR) sensors mounted beneath them, so that their light cannot be kept below the horizontal by tilting them downwards (Fig. 1.55). The effectiveness or otherwise of these units has been covered in Chapter 1.

If the mainstream lighting industry is generally positive about light pollution, why don't they think along

Fig. 3.6. The sky-conscious Total Lighting Exhibition brochure (courtesy Interbuild).

similar lines when making these smaller but often much brighter and more intrusive lights? The answer is twofold. Firstly, the great majority of these lamps are imported from the Far East, and are not made by the large UK- or US-based road lighting companies. Fewer constraints seem to apply, and the few "anti-light-pollution" security lights on the market tend to be made locally. It falls therefore to the *retailers* of such lamps to respond to the prevailing climate, and to use whatever influence they have over the design and manufacture of their suppliers' products to improve the situation. At the time of writing, this is simply not happening, and relatively small numbers of "sky-friendly" home security lamps (Fig. 3.8) sit on retail shelves, looking lost beside great piles of "Rottweilers". The customer, too, should be a regulator of the quality of the product bought, and much modern legislation exists to ensure this. So, education is another part of the answer: if those who believe, rightly or wrongly, that security can be assured by mounting lights on their outside walls know what kind of lights might conceivably have some effect, and, more importantly, what amount of light to use and where to direct it, then the market will be consumer-led.

Earlier, the point was made that minimising light pollution is not about switching off all lights, nor just about correct design of lamp housings to put all the light where it is needed; an important aspect of the solution is putting the *right amount of light* onto the surface to be illuminated, and the lighting industry, with its expertise and advisory capacity, can counsel clients (as many firms already do) as to the minimum wattages for the task, insisting (until legislation can insist for them) that more light is not always good light.

Fig. 3.7. Glare from an indifferently mounted security light, 200 metres away, which forced members of the Wessex Astronomical Society to abandon one of their traditional observing sites in the New Forest.

Fig. 3.8. "Sky-friendlier" security light, illuminating the ground only.

So, the experts, the lighting professionals who are the only ones who can physically solve the problem of light pollution through improved design and the promotion of a climate rewarding good practice, are aware of the problem and have taken more than just first steps towards that solution. CfDS, IDA and other bodies will continue to work alongside them.

What Should Legislators Be Doing About Light Pollution?

Legislators in central government call frequently for the enactment of environmental measures, put their signatures to international environmental agreements such as Agenda 21, wring their hands about the effect that technological progress is having on climate change, and pay lip service to the enforcement of energy-saving directives, but light pollution continues to fall outside the main stream of legislation to control environmental pollution. Indeed, legislators deliberately excluded light from the list of potential pollutants when the British Parliament debated pollution controls in the early 1990s, and the US government has more recently blown hot and cold about its commitment to the reduction of greenhouse gases. Environmental agencies worldwide have committed themselves to international agreements on energy saving and atmospheric and environmental protection; they have accepted targets for these aspects. While recognising and publicising good practice (Fig. 3.9), friends of the night sky in all signatory countries should continue to insist to their elected representatives, from both local and national government, that these targets will include the wasted energy from poor-quality lighting.

Islands of good practice stand out: for example, Michigan's recreational areas have had protection from obtrusive lighting since the success of the Lake Hudson Dark Sky Preserve in the early 1990s. Ontario has a "star park", north-east of Toronto, where nearby communities have agreed to control outdoor lighting. Britain's national parks have a policy of controlling upward light, following the recommendations of the ILE guidance notes (see Appendix 3). But central government seems slow to react: as part of an initiative which spanned the 1990s, the UK Government's Department of the Environment (DoE) targeted schoolchildren as part of an energy-consciousness campaign, through leaflets featuring friendly dinosaurs voicing the slogan "Wasting Energy Costs the Earth". The dinosaurs told the

Fig. 3.9. One government department which has set the trend with well directed lights: the UK Highways Agency's Martin Hazle (left) receives the British Astronomical Association's Good Lighting Award from CfDS committee member Stuart Hawkins (photo: CfDS).

children, and rightly too, that energy-saving light bulbs in the house were a good thing. In spite of entreaties to the DoE by CfDS that a lot of energy was being wasted by environmentally insensitive *exterior* lamps far more powerful than any found on walls or ceilings indoors, the DoE never mentioned in their literature the need to save energy outside the walls. The standard answer from central government in the 1990s was that "education is more likely to lead to a solution of the problem of light pollution than legislation"; and as environmentalists worked to educate, the "Rottweiler" lights were still allowed to appear, bolted to walls everywhere.

Those who can make a difference by regulation and legislation must be made to see that, the longer they choose to ignore the need for positive action on light pollution, the harder it will be for them and for the officers in local government who must carry out their instructions, to repair the damage caused to the environment above; they must acknowledge the value of that part of the environment which has suffered the most damage from human intervention over the last half-century: the starry sky. The actuarial approach seems to be to the forefront of legislators' thinking: light pollution, light trespass and the disappearance of the night sky don't kill people, and a generation brought up beneath orange skies don't seem to complain about what they've never had the chance to see; surely there are forms of pollution which require far more urgent action? However, it is the prime duty of any legislature to ensure the continuation of an acceptable *quality of life* for its citizens, and anyone who has ever had the chance to see the real night sky, and to enjoy the benefits that the darkness of the night can bring to their bodily rhythms and peace of mind, will be able to testify that the loss of the night is an environmental tragedy of some proportions.

Simple but effective measures, on a sensible, evolutionary timescale, to bring back that quality of life would be: the banning of all exterior security lighting whose design and wattage cause glare and emission above the horizontal; an acceleration of the positive trend in well directed roadlighting; real powers for environmental officers to intervene in cases of nuisance by light; and a proper policy of education about light pollution, in parallel with other forms of environmental education.

What Should Local Authorities Be Doing About Light Pollution?

Much of the public lighting we see around us is bought with public money and installed by local authorities. It can be very well done, or *very* badly done (Fig. 3.10), and for our money we can get either a sensitively lit environment with a reasonable chance of seeing the stars, or a garish light-show which defies the heavens.

In the absence of directives from central government, there is still much that a local administration can do to mitigate light pollution if it is so minded. Experience has shown that, with concerned individuals in administrative positions which give them control over relevant decisions, and with the right amount of persistent and friendly pressure from local residents concerned about the environment above, things can happen which will forestall poor lighting.

An early and definitive statement which might serve as a model for other administrations was that of the City of Tucson, Arizona, an astronomically sensitive area with major observatories not far away. Extracts from the Revised Tucson and Pima County Outdoor Lighting Control Ordinances appear below. The complete document is available as Information Sheet 91 on

Fig. 3.10. A local council's poorly directed car park lighting shatters the rural night (photo: Richard Murrin).

the IDA website (see bibliography), and extracts are reproduced with permission.

Ordinance No. 8210.
Tucson/Pima County Outdoor Lighting Code, 21 March 1994.
Section 1. Purpose and Intent. The purpose of this Code is to provide standards for outdoor lighting so that its use does not unreasonably interfere with astronomical observations. It is the intent of this Code to encourage, through the regulation of the types, kinds, construction, installation, and uses of outdoor electrically powered illuminating devices, lighting practices and systems ... (conservation of) energy without decreasing safety, utility, security, and productivity while enhancing nighttime enjoyment of property within the jurisdiction.

Section 2. ...All outdoor electrically powered illuminating devices shall be installed in conformance with the provisions of this Code, the Building Code, the Electrical Code, and the Sign Code of the jurisdiction as applicable and under appropriate permit and inspection.

... Section 4. Definitions.
... Sec. 4.3. "Outdoor light fixture" means outdoor electrically powered illuminating devices, outdoor lighting or reflective surfaces, lamps and similar devices, permanently installed or portable, used for illumination or advertisement. Such devices shall include, but are not limited to search, spot, and flood lights for:
1. buildings and structures; 2. recreational areas; 3. parking lot lighting; 4. landscape lighting; 5. billboards and other signs (advertising or other); 6. street lighting; 7. product display area lighting; 8. building overhangs and open canopies.

... Section 5. Shielding. All nonexempt outdoor lighting fixtures shall have shielding as required by Table 5 of this Code.
Sec. 5.1. "Fully shielded" means outdoor light fixtures shielded or constructed so that no light rays are emitted by the installed fixture at angles above the horizontal plane as certified by a photometric test report.
Sec. 5.2. "Partially shielded" means outdoor light fixtures shielded or constructed so that no more than ten percent of the light rays are emitted by the installed fixture at angles above the horizontal plane as certified by a photometric test report.

... Section 9. Prohibitions.
Sec 9.1. Mercury Vapor Lamps Fixtures and Lamps. The installation, sale, offer for sale, lease or purchase

Table 5. Shielding Requirements [Area A: 35 miles around Kitt Peak National Observatory, 25 miles around Mount Hopkins Observatory; Area B: all area outside Area A outside limits of Indian reservations]

Fixture Lamp Type	Area A Shielded	Area B Shielded
Low pressure sodium[1]	Partially	Partially
High pressure sodium	Prohibited except fully shielded on arterial streets and collector streets of 100 ft or more in right of way width.	Fully
Metal halide	Prohibited[7]	Fully[2,6]
Fluorescent	Fully[3,5]	Fully[3,5]
Quartz[4]	Prohibited	Fully
Incandescent greater than 160 watt	Fully	Fully
Incandescent 160 watt or less	None	None
Any light source of 50 watt or less	None	None
Glass tubes filled with neon, argon, krypton	None	None
Other sources	As approved by the Building Official	

Footnotes:

1. This is the preferred light source to minimise undesirable light emission into the night sky affecting astronomical observations. Fully shielded fixtures are preferred but not required.
2. Metal halide lighting, used primarily for display purposes, shall not be used for security lighting after 11:00 pm or after closing hours if before 11:00 pm. Metal halide lamps shall be installed only in enclosed luminaires.
3. Outdoor advertising signs of the type constructed of translucent materials and wholly illuminated from within do not require shielding. Dark backgrounds with light lettering or symbols are preferred, to minimise detrimental effects. Unless conforming to the above dark background preference, total lamp wattage per property shall be less than 41 watts in Area A.
4. For the purposes of this Code, quartz lamps shall not be considered an incandescent light source.
5. Warm white and natural lamps are preferred to minimise detrimental effects.
6. For filtering requirements for metal halide fixture lamp types see Section 6.
7. Fully shielded and installed metal halide fixtures shall be allowed for applications where the designing engineer deems that color rendering is critical.

of any mercury vapor fixture or lamp for use as outdoor lighting is prohibited.
Sec 9.2. Certain Other Fixtures and Lamps. The installation, sale, offering for sale, lease or purchase of any low pressure sodium, high pressure sodium, metal halide, fluorescent, quartz or incandescent outdoor lighting fixture or lamp the use of which is not allowed by Table is prohibited.
Sec 9.3. Laser Source Light. Except as provided in minor Section 9.4, the use of laser source light or any similar high intensity light for outdoor advertising or entertainment, when projected above the horizontal is prohibited.
Sec 9.4. Searchlights. The operation of searchlights for advertising purposes is prohibited in Area A and is prohibited in unincorporated Pima County. In the territorial limits of the City of Tucson, the operation of searchlights for advertising purposes is prohibited in Area A and is prohibited in Area B between 10:00 p.m. and sunrise the following morning.

Section 10. Special Uses.
Sec 10.1. Recreational Facilities. Any light source permitted by this Code may be used for lighting of outdoor recreational facilities (public or private), such as, but not limited to, football fields, soccer fields, baseball fields, softball fields, tennis courts, auto race tracks, horse race tracks or show areas, provided all of the following conditions are met:
a. Lighting for parking lots and other areas surrounding the playing field, court, or track shall comply with this Code for lighting in the specific Area as defined in Section 4.4 and 4.5 of this Code.
b. All fixtures used for event lighting shall be fully shielded as defined in Section 5 of this Code, or be designed or provided with sharp cut-off capability, so as to minimise up-light, spill-light, and glare.
c. All events shall be scheduled so as to complete all activity before or as near to 10:30 p.m. as practical, but under no circumstances shall any illumination of the playing field, court, or track be permitted after 11:00 p.m. except to conclude a scheduled event that was in progress before 11:00 p.m. and circumstances prevented concluding before 11:00 p.m.
Exception: (City only.) Any portion of a recreational facility located within 300 feet of a road or street designated as a scenic route shall be lighted using only fixtures approved for use under this Code for the Area, as defined in Section 4.4 and 4.5 of this Code, in which said recreational facility is located.
Exception: (County only.) Recreational facilities located along roads and streets designated as scenic

routes shall be lighted using only fixtures approved for the Area in which they are located.

Sec. 10.2. Outdoor Display Lots. Any light source permitted by this Code may be used for lighting of outdoor display lots such as, but not limited to, automobile sales or rental, recreational vehicle sales, or building material sales, provided all of the following conditions are met:

a. Lighting for parking lots and other areas surrounding the display lot shall comply with this Code for lighting in the specific area as defined in Section 4.4 and 4.5 of this Code.

b. All fixtures used for display lighting shall be fully shielded as defined in Section 5 of this Code, or be designed or provided with sharp cut-off capability, so as to minimise up-light, spill-light, or glare.

c. Display lot lighting shall be turned off within thirty minutes after closing of the business. Under no circumstances shall the full illumination of the lot be permitted after 11:00 p.m. Any lighting used after 11:00 p.m. shall be used as security lighting.

... Section 12. Other Exemptions.

Sec. 12.1. Nonconformance

1. Mercury vapor lamps in use for outdoor lighting on the effective date of the ordinance codified in this chapter shall not be so used.

2. (City) Bottom-mounted outdoor advertising sign lighting shall not be used. (County) Bottom-mounted outdoor advertising sign lighting shall not be used, except as provided in Section 7.

3. All other outdoor light fixtures lawfully installed prior to and operable on the effective date of the ordinance codified in this chapter are exempt from all requirements of this Code except those regulated in Section 7 and in minor Sections 9.3 and 9.4 and in Section 10. There shall be no change in use or lamp type, or any replacement or structural alteration made, without conforming to all applicable requirements of this Code.

Sec. 12.2. Fossil Fuel Light. All outdoor light fixtures producing light directly by the combustion of natural gas or other fossil fuels are exempt from all requirements of this Code.

Sec. 12.3. State and Federal Facilities. Outdoor light fixtures installed on, and in connection with those facilities and land owned or operated by the federal government or the state of Arizona, or any department, division, agency or instrumentality thereof, are exempt from all requirements of this Code. Voluntary compliance with the intent of this Code at those facilities is encouraged.

> ... Section 15. Violation.
> It shall be a civil infraction for any person to violate any of the provisions of this Code. Each and every day during which the violation continues shall constitute a separate offense.
>
> Section 16. Enforcement and Penalty.
> Sec. 16.1. [City only] Pursuant to Section 28-12 of the Tucson Code:
> 1. When a violation of this Code is determined, the following penalty shall be imposed:
> a. A fine of not less than fifty dollars nor more than one thousand dollars per violation. The imposition of a fine under this Code shall not be suspended.
> b. Any other order deemed necessary in the discretion of the hearing officer, including correction or abatement of the violation.
> 2. Failure of a defendant to comply with any order contained in a judgment under this Code shall result in an additional fine of not less than fifty dollars nor more than one thousand dollars for each day the defendant fails to comply.
> Sec. 16.1. [County only] A violation of this Code is considered a civil infraction. Civil infractions shall be enforced through the hearing officer procedure provided by A.R.S. Section 11-808 and Sections 18.95.030, 18.95.040, and 18.101.60 of this Code [The numbering scheme of the Sections is different in the County Code]. A fine shall be imposed of not less than fifty dollars nor more than seven hundred dollars for any individual or ten thousand dollars for any corporation, association, or other legal entity for each offense. The imposition of a fine under this Code shall not be suspended.

Many Districts and Boroughs in the UK have adopted "Light Pollution" or "Light Trespass" clauses into their updated local plans (Fig. 3.11). This helps the planning departments of local authorities to ensure that good-quality external lighting schemes are incorporated into plans at approval stage. Poorly designed or overbright schemes can be referred back to the applicant for modification. In this way, poor lighting schemes are nipped in the bud *before* they become a problem.

Here are extracts from adopted local plans which might serve as examples of good practice to local councils in any country.

> Policy E6 Swale Borough Local Plan 1994
> The Borough Council will seek to minimise light pollution. Details of any lighting scheme required as part of any new development should be submitted as part of

Fig. 3.11. East Dorset District Council have light pollution on their agenda: leader Don Wallace receives the Good Lighting Award (photo EDDC).

the planning application. Applicants will be expected to demonstrate to the local planning authority that the scheme proposed is the minimum needed for security and working purposes and that it minimises potential pollution from glare and spillage, particularly to:
(1) residential and commercial areas; areas of nature conservation interest; and areas whose open and remote landscape qualities would be affected.

East Hampshire District Council Local Plan
Details of any external lighting scheme required as part of any new development should be submitted as part of the planning application. In order to minimise light pollution and increase energy efficiency, the District Council will need to be satisfied that the lighting scheme proposed is the minimum required for security and working purposes and that it minimises potential pollution from glow and spillage. On the edge of settlements and in rural locations, landscaping measures should be provided to screen the lighting installation from view. Conditions will be attached to any floodlighting approvals given for evening usage of sport facilities such as pitches, tennis courts and golf driving ranges to control light intensity, light spillage and hours of use.

Environmental Policy 6, Malvern Hills Local Plan
Applications for development requiring or likely to require external lighting shall normally include details of lighting schemes which will be expected to demonstrate that: the lighting scheme proposed is the minimum required to undertake the task, light spillage is minimised in the edge of town or village locations,

or in rural areas, landscaping measures will be provided to screen the lighting installation from view from the neighbouring countryside areas, and there will be no dazzling or distraction of drivers using nearby highways.

Hinckley and Bosworth Borough Council
Light pollution is caused by excessive artificial light being directed into the night sky. Outdoor lighting can cause intrusive and unnecessary pollution in both urban and rural areas, although it is in the countryside that light pollution is most noticeable. Excessive light in the night sky is visually intrusive and is also a significant waste of energy. The visibility of the stars is much reduced by light pollution. It is therefore important in the interests of visual amenity and energy conservation that light pollution is prevented and where possible reduced. Through good design of lighting, the reduction of light pollution should not conflict with the principles of crime prevention and safety.

Epsom and Ewell Local Plan
Artificial light is increasingly being perceived as a form of pollution. Illuminated advertisements, floodlit sports facilities, security lights and street lights can all contribute to pollution such as skyglow and glare. They can damage visual amenity, disturb people's sleep, and affect local ecology. Planning control over artificial light other than advertisements is generally limited to new structures or works which are integral to other development. However, where planning permission for artificial light sources is required the Council will seek to prevent detrimental impact on surrounding areas. Impact will be minimised by ensuring that artificial light is carefully sited, appropriately shielded, directed only onto the specific area where it is needed, and designed at the minimum height and brightness to serve its purpose. Where appropriate, the Council will use conditions to limit the hours of illumination. Developers' attention is drawn to the Institute of Lighting Engineers' publication "Guidance notes for the reduction of light pollution".

Further measures which local administrations might take could be to survey local residents on schemes whose visual impact would affect large numbers of them, such as the long-term floodlighting of public buildings and churches (Fig. 3.12), to see whether the schemes have widespread support from those who might conceivably be contributing to the cost. Several administrations have already issued pamphlets to all residents, defining the problem and incorporating guidelines on reducing light pollution.

Fig. 3.12. Some church floodlighting is far too random, and allows a large fraction of emissions into the sky (photo: Chris Baddiley).

What Should Architects Be Doing About Light Pollution?

"More nightmare than nightscape": this is how our overlit and garish cities were described in the Royal Fine Art Commission's booklet *Lighten our Darkness*, published in 1993. As a design concept, the lighting of structures, and the actual lamps mounted on them for both daytime decorative effect or night-time floodlighting, are often viewed, it seems, in isolation from the wider surroundings; worse still, clients ordering lighting schemes might, as lighting professionals warned at a meeting on the subject of lighting and the environment in Glasgow in 1999, try to outdo neighbouring schemes in brightness and general effect. The result is a garish, "more nightmare than nightscape" scene. Individual elements of schemes are often overstated, in an attempt to "paint over" light falling upon them from local street lights or neighbouring structures. The architectural merit of many buildings, far from being enhanced by floodlighting, is lost because the lights draw attention to themselves, rather than to what is being lit, with too-vivid colours or simply far too much power (Fig. 3.13). For buildings to stand out from the darkness, there has to be some darkness present. Indeed, do some of the buildings washed with light in

Fig. 3.13. A city nightscape, with buildings clamouring for attention.

our towns need to be lit at all? Hal Moggridge, Commissioner of the Royal Fine Art Commission, wrote in 1993: "Too much light can seem superfluous, no more than a symbol of waste of energy. Lighting should not be seen as any more inherently desirable than darkness."

Rural buildings may often be lit to draw attention to their architectural features at night. Does a certain Scottish castle look better flushed an unnatural green, the whole reflected in the nearby loch, or as a stark, dark shape picked out by moonlight against the inky black background of a pine forest? Does an isolated country church have more merit floodlit by orange lights concreted to the ground around it, reducing it to a bland monochrome façade and dazzling anyone passing through the churchyard at night, or would it look better with a bright white light source inside in the evening, projecting the beauty of the stained-glass windows, with perhaps a narrow-beam spotlight or two outside carefully trained on features which deserve attention? Such value judgements do not really further the general debate about light pollution, since one man's meat is another's poison; but architects and design consultants can certainly exercise responsibility towards the environment both around and above the structures they wish to floodlight. Many consultancies already have the environmentalists' and professionals' concerns at heart when deciding on lighting strategies for new developments. In *Edinburgh Lighting Vision*, issued by consultants Lighting Design Partnership, we read: "Artificial light is a powerful tool, and needs to be used sensitively, selectively and imaginatively, whether lighting a building, a street, a garden or a monument. Misuse results in visual distortion, incorrect visual information and, in the wider context, an imbalance between the numerous aspects which comprise a cityscape at night." Carefully prepared schemes and a general consensus on what is needed to tone down the urban "nightmare" could be the result of a proper code of practice on lighting among architects and consultancies: a code within which lights are not seen just as decorative adjuncts during the day, and a means of drawing attention away at night from all else and towards the structure on which they stand. The watchwords informing this code should be *selectivity* and *collaboration*: select the right buildings, the right times for them to be lit and the right times for them not to be lit, and the most appropriate, most environmentally sensitive lights, if schemes are to go ahead (in some British villages, church floodlighting schemes have been voted down, and many of the schemes which have gone ahead have done so without public consultation); collaboration is necessary with local residents, environmentalists, and most importantly, with other architects and consultants, through the channel of a local authority or chamber of commerce, to ensure that both the built and the natural environments are harmonised and respected, and townscapes enhanced rather than vandalised, and, last but not least, that those directly affected by proposed lighting schemes should have their say before any decision is made.

What Should Retailers Be Doing About Light Pollution?

Much of the responsibility for the current proliferation of poor-quality, overbright security lighting lies with retailers, who claim blandly that such units will deter crime; they are imported in large numbers and usually sold very cheaply. Many retailers do not readily respond to information put before them by astronomers, environmentalists and by some lighting professionals, who agree that such lights are, in the main, insufficient and environmentally insensitive, and although units with better light direction are finding their way onto the market, this is not accompanied by a decline in the promotion of less sensitive types. A well-known retailer claims that all its lines are environmentally sound. While the origin of its timber products and the non-toxicity of its paints may well be praiseworthy, the range of exterior lights offered represents a "blind spot" in this retailer's environmental credentials. Tighter specifications, including better shielding, sensible wattages and information in packaging, are surely not beyond the means of the DIY business, even if it adds a little to the price. What price the environment they claim to be protecting?

What Should Astronomers Be Doing About Light Pollution?

It is through astronomers, both amateur and professional, that the various threads of the solution will be drawn together. They may not be designers or sellers of lamps, they may not manufacture them, and they may not belong to local administrations (although some do, and have had great success in influencing the choice of lighting), but what they can do is, individually or in groups, to make sure that all the interests mentioned here, all those who make, choose and install lights, know about the dark skies debate and reflect on it before those lights are made, chosen or installed. The prevailing mood among most astronomers, that the problem is just too far advanced and that their voices will not be heard against the "noise" of traditional assumptions about light, and powerful vested interests, is in error. IDA and CfDS have proved that, from small beginnings, the dark skies message can be spread very widely. They know that the various groups involved *will* listen, and that changes in lighting *can* come about, slowly but surely. "Great oaks from little acorns grow." The astronomers' task is to accelerate those changes by publicising the issue frequently, both as individuals and *en masse*. We, the astronomers, are never going to get all lights switched off, and most of us would not want to: we have the same lighting needs as everybody else. Enforceable controls over lighting, and a sane approach to its installation and positioning, *are* achievable goals.

A good way, if you are an amateur astronomer, to start moving towards those goals on a very local scale is to approach individuals and organisations with offending lights in your immediate area and point out that it could be much better done. I know from having spoken to hundreds of astronomers about neighbours' poor lights that many of them do not care to (politely!) confront offenders: they would be surprised at how often such approaches work – those causing light pollution usually do not even realise that they are causing a problem, and they will take action.

On a slightly less local scale, find out who in your area is concerned about the spread of "over-the-top" lighting, with a view to forming an active pressure group. Use local media to air your views: they welcome "green" issues. Offer to talk to local groups who use regular speakers, and hold special public meetings of your astronomical society, at a dark site if you have one, to show people what they are missing in poorly lit urban areas. You may well know of a rural site from which the depredations of light pollution are very easily demonstrated, with a view of a distant town's myriad lights and the halo of waste above them blotting out the stars for many degrees above the horizon, while the Milky Way remains tantalisingly visible vertically above: invite local councillors, journalists, teachers, environmentalists, and owners of poor lighting to these places to make the comparison between starry and glary skies. An explanation of the problem of light pollution at such a site may be worth many letters to local papers or talks to interest groups. You will be surprised at how many people care: most of them non-astronomers, and many of them of the older generation who can remember what the night sky and a comfortable terrestrial nightscape should look like, and realise what they have lost.

Next, organise letters and petitions (see Appendix 6) to local authorities from your supporters. Find out who is responsible within your local administration for choosing and installing outside lighting, and write to named individuals, asking them their views on the problem and what they are already doing about it. Quote the ordinance and local plans shown above in the "local authorities" section, and the advice from IDA and CfDS.

A positive approach and politeness usually get results. "Broadsides" don't win friends, carping criticism builds barriers, and baldly accusing someone of being a polluter is not wise. Have the facts to hand: published engineers' guidelines, factsheets from IDA,

CfDS and other organisations listed in Appendix 1, and findings such as those in Parts 1 and 2, on lighting and crime and the cost of wasting energy skywards. Join IDA and CfDS to benefit directly from their updates, literature and support, and refer to their websites.

Public interest in astronomy is growing. Many astronomical societies reported an influx of new members after the passage of Comet Hale-Bopp in 1997: recent eclipses, space station sightings and the growing profile of the "Spaceguard" movement (who doesn't now know what the consequences of a too-close encounter with a 1 km asteroid would be?), all these have resensitised many people to the fact that there's a universe out there. Astronomers should ride this tide, and stress in whatever ways they can that this rediscovered environment could offer so much more if correctly protected.

Your over-lit neighbours may be sceptical on hearing you argue for light control. Surely, they will say, a well-lit neighbourhood is essential in these crime-ridden days? Now's the time to show them, "in the field", what "well-lit" means. Try three simple demonstrations which might clear their minds; they have certainly worked with people in my area.

- Ask neighbours with a blinding, outward-facing security light to stand in the street while you walk towards their property, disappearing behind the wall of light it emits. They will realise that theirs is really an "anti-light", concealing rather than revealing, and they may well be moved to do something about it.
- If new, more downward-directed lamps are installed near your observing site, people may complain that the street lights have lost their "sparkle" (they mean glare) and are dimmer, even though they may be of the same or even higher wattage than those they replace, and light the road more efficiently. Prove the effectiveness of the new lights by asking doubters to stand with you half-way between the columns, and ask them to read the 1-millimetre-high print on one of their credit cards. It is usually possible to do this, and difficult for anyone attending this demonstration still to maintain that it's dark between the lamps.
- Many neighbours have re-angled or re-sited lamps after looking through an astronomer's telescope: occasional star parties will sensitise your neighbours to the existence and value of the environment above (Fig. 3.14). Few will not be moved by their first view

Fig. 3.14. Preparing for a garden star party.

of the Perseus Double Cluster: "like opening a jewel box", said my neighbour, who has installed a switch on her garden light and takes care to turn it off in the late evening. The rings of Saturn seen for the first time so impressed another person in my road that he fitted a lower-wattage bulb to his porch light to cut down glow generally in the neighbourhood. Perhaps you can even arrange for offending security lights to be triggered while their owner is stargazing with you, so that the effect can be appreciated first-hand!

You may not be able to convince all your neighbours, especially in high-crime areas, that a dark environment can be as much of a deterrent to wrongdoers as a brightly lit one (see the section above on "Security Lamps and Crime"), but you can certainly make the point that a real human being outside at night is a far better security device than any lamp: an astronomer can take action if (s)he sees or hears anything suspicious, but a security light cannot.

Perhaps the greatest problem that concerned astronomers face is the gut reaction, which will sometimes come from even the most educated listener, that they are against lighting, and accusations will fly from groups and individuals who, rightly or wrongly, feel vulnerable when going about their legitimate business at night. So, introduce the fact that good-quality lighting means a more evenly lit environment early on in any presentation or discussion; discuss the demerits of glare, and the possibilities of concealment through glare or too-deep shadows. List the security and other benefits of a properly lit terrestrial environment before you move on to the astronomical arguments. It is vital to remember that not everybody considers that the night sky is an important thing, and they have a perfect right if they wish to have different priorities and other interests and leisure pursuits. Discuss the savings in energy and money which good lighting brings, since saving money and energy have far greater relevance than astronomy in most lives. Talk about the use and misuse of a security light, and only then introduce non-astronomers to the beauty and value of the stars with

Fig. 3.15. I wish you clear skies.

some of the superb images which you or fellow astronomers have taken, or which are on offer from many sources; stress that the ultimate solutions to light pollution mean that, in the words of David Crawford: "everybody wins".

Every astronomer, amateur or professional, is potentially an interchange point in an enormous network of information about the problem of skyglow and how it may be addressed. Talk to anyone and everyone about it – perhaps not too often: a skyglow bore is just as bad as any other bore – and be part of that network. Support the work of CfDS, IDA and other campaigners.

Light pollution is not a problem which we will solve at a stroke, and those working to turn the tide at the beginning of the new century should perhaps resign themselves to the fact that not they, but their descendants, will be able to see a truly worthwhile night sky from urban areas and from a sensitively lit countryside – this is an issue for the altruistic and patient! I believe that darker skies will come to many of us in the twenty-first century (Fig. 3.15): exactly when that desire becomes a general reality depends very much on how many of us make a noise about light-energy waste, and our persistence in doing so. I wish you clear skies.

Appendix 1
Information about Organisations Committed to Reducing Light Pollution

Below are details of organisations of which the author has heard whose main thrust involves countering the effects of poor lighting practice on the night sky, or who have published guidelines or charters on the subject:

Australia

IDA Section: Reg Wilson
Lighting Analysis & Design
32 Carina Rd.
Turramurra,
NSW 2074
Australia.

Phone 61-2-488-7078
Fax 61-2-9488-7078
E-mail: regrw@acay.com.au

Sydney Outdoor Lighting Improvement Society
PO Box 999
North Turramurra
NSW 2074
Australia.

E-mail: solislp@netscape.net

Belgium

VVS Werkgroep Lichthinder
E. Miroen
Volkssterrenwacht Urania
Jozef Mattheessenstraat 60
B2540 Hove
Belgium

Canada

IDA Section: Rob Dick
Light Pollution Committee
Canadian Campaign for Dark Skies
Royal Astronomical Society of Canada
2 Forest Laneway, Apt. 2409
Toronto,
ON M2N 5X9
Canada.

Phone 416-924-9639
E-mail: rasc@rasc.ca
Web: www.rasc.ca/light/

IDA Alberta section
Edmonton Centre
Royal Astronomical Society of Canada
Attn: Howard Gibbins
c/o 6911-98A Avenue
Edmonton,
AB T6A 0B9
Canada.

Phone 403-469-9765
E-mail: howardg@ibm.net

Denmark

Astronomisk Selskab
Per T. Aldrich
Naesbyholm 6 st.th.
2700 Bronshoj
Denmark.

E-mail: paldrich@inet.uni-c.dk

Finland

IDA Finland
URSA Astronomical Society
Laivanvarustajankatu 9 C 5
SF-1708 Helsinki
Finland.

E-mail: Mika.Pirttivaara@ursa.fi

France

AFA (Association Française d'Astronomie)
Light Pollution section
Eric Fourlon

E-mail: Eric.Fourlon@cnrs-dir.fr

Comité pour la Protection du Ciel Nocturne
Laurent Corp
56 Avenue de Paris
12000 Rodez
France.

Phone +33-5-65-78-58-61
E-mail: lcorp@wanadoo.fr
Web: www.astrosurf.org/lcorp

FAPAM (Fédération d'Astronomie Populaire du Midi)
Jean-Marie Lopez
La Babote
Boulevard de l'Observatoire
34000 Montpellier
France.

Association Nationale pour la Protection du Ciel Nocturne
Alain Legué

E-mail: Alain.Legue@wanadoo.fr

Association Andromède Astronomie Aveyronnaise (4A)

Web: www.astrosurf.org/mercure/andromede

Germany

Dark Sky
VdS (Vereinigung der Sternfreunde)
Wolfgang M. Wettlaufer
Weissdornweg 14/37
D-72076 Tuebingen
Germany.

E-mail: aaptue@ait.physik.uni-tuebingen

Greece

Dr Margarita Metaxa
63 Ethnikis Antistaseos Str
Athens 152 31
Greece.

Phone: 301-674-28-25
E-mail: mmetaxa@compulink.gr
Web (English version) www.epolian.gr/LP/lp.htm
Web (Greek version) www.epolian.gr/LP/grlp.htm

Italy

Dr Mario Di Soro
Osser. Astron. Di Campr. Catino
Via Fosse Ardeatine 234
03100 Frosinone
Italy.

E-mail: mario.disora@rtmol.stt.it

Japan

Shigemi Uchida
4-21-206 Wakabadai Asahiku
Yokahama 241–0801
Japan.

Phone +81-45-921-2334
E-mail: suchida@mvb.biglobe.ne.jp
Web: www2a.biglobe.ne.jp/~wakaba/

Malta

Light Pollution Awareness Group
Astronomical Society of Malta
Alexei Pace
25 Triq Santa Katarina
Marsaxlokk ZTN 09
Malta.

E-mail: alexei@e-architect.com
Web: maltamedia.com/astro

Slovenia

IDA Slovenia
Dr Herman Mikuz
University of Ljubljana
Department of Physics
Astronomical Observatory
Pot na Golovec 25
1000 Ljubljana
Slovenia

Phone: +386 1 501353
Fax: +386 1 505370
E-mail: herman.mikuz@uni-lj.si

Dark Sky Slovenia
Ljubljana
Slovenia
Web: www.fiz.uni-lj.si/astro/comets/DSSi/index.html
E-mail: herman.mikuz@uni-lj.si

South Africa

Cliff Turk
Astronomical Society of Southern Africa
(Cederberg Observatory)
20 Nerine Ave
Pinelands
7405 South Africa.

E-mail: cliffturk@yebo.co.za
Tel: +27 (0)21 531-5250

Spain

Oficina Tecnica Para la Proteccion de la Calidad del Cielo (OTPC)
Instituto de Astrofisica de Canarias
38200 La Laguna
Tenerife
Canary Islands

E-mail: fdc@iac.es / fpaz@iac.es
Web: www.iac.es/proyect/otpc/otpc.htm

Grupo de cielo oscuro de la Agrupación Astronómica de Madrid
Web: www.iac.es/AA/AAM/oscuro.html

Collectiu Cel Fosc
Web: www.celfosc.org

Iniciativa aragonesa para el control de la contaminacio luminica
Web: www.astrored.org/doc/luz/iniciativa.html

La contaminació lumínica a Catalunya
Web: www.am.ub.es/contaminacio-luminica/cl.html

Switzerland

Dark-Sky Switzerland (DSS)
Philipp Heck
Postfach, CH-8712 Stˇfa
Switzerland.

E-mail: info@darksky.ch
Web: www.darksky.ch

United Kingdom

The British Astronomical Association: Campaign for Dark Skies (CfDS).
BAA, Burlington House,
Piccadilly, London W1J 0DU
England.

Phone: 0207 734 4145
E-mail: office@baahq.demon.co.uk
Web: www.dark-skies.freeserve.co.uk

Council for the Protection of Rural England (CPRE).
CPRE, Warwick House,
Buckingham Palace Road
London SW1W 0PP
England.

Phone: 0207 976 6433
Web: www.greenchannel.com.cpre

Institution of Lighting Engineers (ILE): see bibliography.

Scottish Natural Heritage
2 Anderson Place
Edinburgh EH6 5NP
Scotland.

Tel: 0131 446 2405

USA

Illuminating Engineering Society of North America (IESNA): see bibliography.

International Dark-Sky Association (IDA)
3225 North First Avenue
Tucson AZ 85719
USA.

Phone: 520-293-3198
E-mail: ida@darksky.org
Web: www.darksky.org

To receive the quarterly IDA newsletter, contact IDA at the address above.

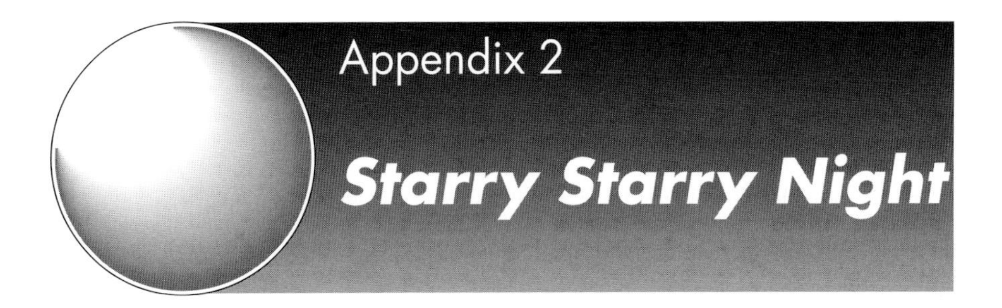

Appendix 2

Starry Starry Night

In 1993, The British Astronomical Association's Campaign for Dark Skies (CfDS) joined forces with the Council for the Protection of Rural England (CPRE) to produce the widely distributed and much quoted booklet *Starry Starry Night*, revised in 2000. Its contents may be of use to anyone seeking to combat rural light waste, and the bulk of its text is reproduced here.

Introduction

Dark skies, moonlit and star-studded nights and the dawning of the day are part of the rich variety of our countryside. Yet nowadays much of the country is lit, often throughout the night. With more and more road lights, security lights and floodlights, we are in danger of losing our starlit skies, twilight and the emerging daybreak.

Outdoor lighting can cause intrusive and unnecessary pollution of our countryside. Poorly designed or badly aimed lights are responsible for "skyglow". This scattered light spills into and colours the night sky and reduces the visibility of the stars. Illuminated skies blur the separation between country and town. They reduce the feeling of remoteness in rural areas and introduce a suburban character deep into the countryside.

The Council for the Protection of Rural England and the British Astronomical Association are pressing for:

- better protection for our remaining unlit landscapes and countryside;
- greater attention to the siting and type of lighting used both in the country and in towns, in order to reduce wasted light; and

- removal of unnecessary lighting because of its impact on the night sky.

Our countryside illustrates natural patterns and tones. In daylight these are patterns of colour and texture – shades of green or lines of the landscape. At night the patterns are of light – starlight, moonlight and shadow. But we are losing the splendour of our starlit skies and moonlight shadows as more and more of our surroundings are illuminated throughout the night.

The astronomers' plea to darken our skies is a warning that unless this situation is reversed, the view of the stars will be lost. With it we would lose the majestic beauty of the night skies and the scientific and cultural inspiration this has provided since ancient times.

Problems from Lighting

We are using more and more lights in development and along roads, and our lighting is becoming increasingly bright. Lighting is needed in many areas in the interests of public safety and it can enhance the appearance of some public buildings. But a lot of external lighting is poorly designed and misdirected. Some is unnecessary. Much is wasteful and intrusive.

Particular problems arise from:

- new, poorly designed street lighting in small villages;
- increasing ribbons of road lights cutting through our countryside;
- illuminated shop windows and advertising signs left on overnight;
- badly designed lighting in car parks, stations and shopping centres;
- over-bright domestic security lighting flooding the neighbourhood in bright light;
- badly floodlit sports facilities, such as golf driving ranges, or motorway service areas, which bathe rural areas in brightness; and
- new housing estates or shopping complexes with discordant lighting, often much more intrusive than neighbouring lighting.

Safety and security may not always be improved by lighting. Bright lights are not necessarily more effective than a low level of light. Dazzling lights cause deep shadows which can compromise security around warehouses, car parks and homes.

Action

CPRE and BAA want action now. We need:

- Government planning guidance on when and how to control lighting in order to reduce light pollution and energy waste;
- local authority planning policies which protect unlit landscapes and countryside and which control lighting in new development;
- light pollution to be recognised as a statutory nuisance similar to noise pollution;
- full assessment of lighting proposals in roads and other development schemes;
- local authorities to review the impact of existing lighting and with lighting engineers to put forward schemes to reduce this impact;
- recognition by the Government that mechanisms other than lighting have a rôle in reducing night-time accidents;
- guidance for planners, highway authorities and developers on the most efficient and effective lighting systems, in particular systems which limit upward light;
- information for the public on minimising intrusive lighting through the use of low-intensity, sensor- or time-controlled and well directed domestic lighting systems; and
- all new domestic outdoor lighting systems to include information on correct installation to minimise light pollution.

What you can do:

- Press for planning policies to protect unlit countryside and to reduce light pollution.
- Ask for lighting details to be included in individual planning applications.
- Check whether road schemes include lighting proposals, and seek public consultation over proposals to light existing roads.
- Make your contribution by reducing light pollution where you live.
- Campaign to protect our remaining dark landscapes.

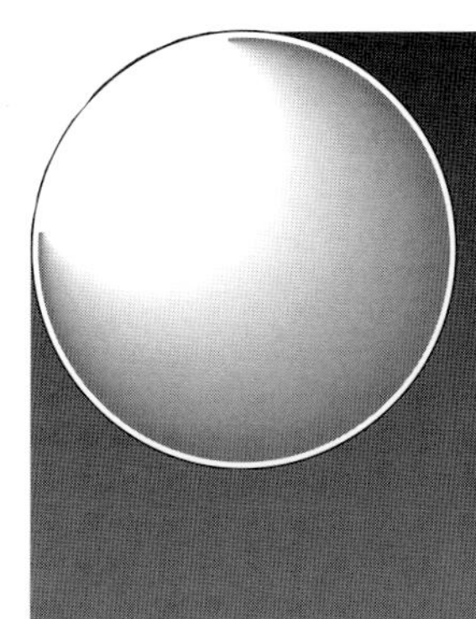

Appendix 3

Recommendations for Good Light Control: the ILE Guidance Notes and the NELPAG/IDA *Good Neighbor Outdoor Lighting Advice*

ILE Guidance Notes

The Institution of Lighting Engineers (see bibliography) has produced the following guidelines for good lighting control (reproduced with permission):

Guidance Notes for the Reduction of Light Pollution

All living things adjust their behaviour according to natural light. Man's invention of artificial light has done much to safeguard and enhance our night-time environment but, if not properly controlled, obtrusive light (commonly referred to as light pollution) can present serious physiological and ecological problems. Light pollution, whether it keeps you awake through a bedroom window or impedes your view of the night sky, is a form of pollution and could be substantially reduced without detriment to the lighting task.

Skyglow, the brightening of the night sky above our towns and cities; glare, the uncomfortable brightness of a light source when viewed against a dark background; and light trespass, the spilling of light beyond the boundary of the property on which the light source is located, are all forms of obtrusive light. This is not only a nuisance, it wastes electricity and thereby large sums of money, but more importantly it helps destroy the Earth's finite energy resources, resulting in unnecessary emissions of greenhouse gases.

Listed below are some easy ways to reduce the problems of unnecessary, obtrusive light:

[Al] Do not "over" light. This is a major cause of light pollution and is a waste of money. There are published standards for most lighting tasks.

[A2] Switch off lights when not required for safety, security or enhancement of the night-time scene. In this respect one can introduce the concept of a curfew, i.e. a period in which more restrictive controls are applied to obtrusive light. In all new developments there is scope for Local Planning Authorities (LPAs) to impose conditions relating to curfew hours in determining planning applications. For instance the LPA may determine that non-essential lighting, such as advertising and decorative floodlighting, should be switched off between 23.00 hours and dawn. In the case of new non-residential developments LPAs are encouraged to impose such curfews. The attachment of domestic security and decorative lighting to residential buildings often does not require planning permission. However, as the floodlights are operational throughout the night, it is considered that the after curfew levels of lighting control shown in Table 1 should be used at all times.

[A3] Use specifically designed lighting equipment that minimises the upward spread of light near to, or above the horizontal. Care should be taken when selecting luminaires to ensure that the units chosen will reduce spill light and glare to a minimum. The use of luminaires with double-asymmetric beams designed so that the front glazing is kept at, or near parallel to, the surface being lit will assist in the reduction of glare, provided the units are correctly aimed. Similarly, modern well-controlled projector type luminaires, which can be aimed very precisely, can give an excellent cut-off beyond the lit area so reducing spill light and glare.

[A4] Keep glare to a minimum by ensuring that the main beam angle of all lights directed towards any potential observer is kept below 70 degrees. Higher mounting heights allow lower main beam angles, which can assist in reducing glare. In areas with low ambient lighting levels, glare can be very obtrusive and extra care should be taken when positioning and aiming lighting equipment. When lighting vertical structures such as advertising signs, direct light downwards, wherever possible, to illuminate them; not upwards. If there is

Table 1. Obtrusive light limitations for exterior lighting installations

Environmental zone	Skyglow ULR (max.%)	Light into windows E_v [Lux] (1)		Source intensity I [kcd] (2)		Building luminance before curfew (3)	
		Before curfew	After curfew	Before curfew	After curfew	Average L[cd/m²]	Maximum L [cd/m²]
E1	0	2	1*	0	0	0	0
E2	2.5	5	1	20	0.5	5	10
E3	5.0	10	2	30	1.0	10	60
E4	15.0	25	5	30	2.5	25	150

* Acceptable from public road lighting installations ONLY.

Where:
ULR = Upward Light Ratio of the Installation and is the maximum permitted percentage of luminaire flux for the total installation that goes directly into the sky. (formerly UWLR)
Ev = Vertical Illuminance in Lux normal to glazing
I = Light Intensity in Candelas
L = Luminance in Candelas per Square Metre

Notes:
LIGHT INTO WINDOWS – These values are suggested maximums and need to take account of existing light trespass at the point of measurement.
SOURCE INTENSITY – This applies to each source in the potentially obtrusive direction, outside of the area being lit. The figures given are for general guidance only and for some large sports lighting applications with limited mounting heights, may be difficult to achieve. If the aforementioned recommendations are followed then it should be possible to further lower these figures.
BUILDING LUMINANCE – This should be limited to avoid over lighting, and relate to the general district brightness. In this reference building luminance is applicable to buildings directly illuminated as a night-time feature as against the illumination of a building caused by spill light from adjacent floodlights or floodlights fixed to the building but used to light an adjacent area.
These limitations may be supplemented by a Local Planning Authority's own planning guidance for exterior lighting installations and you are therefore recommended to check with the Local Planning Authority before designing or installing any exterior lighting.

no alternative to uplighting, then the use of shields, baffles and louvres will help reduce spill light around and over the structure to a minimum.

[A5] For road lighting installations, light near to and above the horizontal should be minimised to reduce glare and visual intrusion (Note ULRs in Table 1). The use of full horizontal cut-off luminaires installed at 0 degrees uplift will minimise visual intrusion within the landscape as well as upward light. However, in many urban locations, luminaires fitted with a shallow bowl providing good control of light near to and above the horizontal can provide a satisfactory solution whilst maximising the spacing of the luminaires.

Environmental Zones

It is recommended that the Local Planning Authority as part of their Development Plan specify the following environmental zones for exterior lighting control.

Category	Examples
E1:	Intrinsically dark areas: National Parks, Areas of Outstanding Natural Beauty, etc.
E2:	Low district brightness areas: Rural or small village locations
E3:	Medium district brightness areas: Small town centres or urban locations
E4:	High district brightness areas: Town/city centres with high levels of night-time activity

Where an area to be lit lies on the boundary of two zones or can be observed from another zone, the obtrusive light limitation values used should be those applicable to the most rigorous zone.

NELPAG/IDA *Good Neighbor Outdoor Lighting*

The New England Light Pollution Advisory Group, in association with the International Dark-Sky Association, offers the following general advice (reproduced with permission):

Why is there outdoor lighting?

Outdoor lighting is used to illuminate roadways, parking lots, yards, sidewalks, public meeting areas, signs, worksites and buildings. It provides us with better visibility and a sense of security.

When well designed and properly installed, outdoor lighting can be and is very useful in improving visibility and safety and a sense of security, while at the same time minimising energy use and operating costs.

Why should we be concerned?

If outdoor lighting is not well designed and properly installed, it can be costly, inefficient, glary, and harmful to the night-time environment. These are the issues:

Glare: poorly designed or poorly installed lighting can cause a great deal of glare that can severely hamper the vision of pedestrians, cyclists and drivers, creating a hazard rather than increasing safety. Glare occurs when you see light directly from the fixture (or bulb);

Light trespass: poor outdoor lighting shines onto neighborhood properties and into bedroom windows, reducing privacy, hindering sleep, and creating an unattractive look to the area;

Energy waste: much of our outdoor lighting wastes energy because it is not well designed. This waste results in high operating costs and increased environmental pollution from the extra power generation needs. We waste over a billion dollars a year in the United States alone, lighting up the sky at night;

Skyglow: a large fraction of poor lighting shines directly upwards, creating the adverse skyglow above our cities that washes our view of the dark night sky, taking away an important natural resource. In addition to the cost savings, less skyglow will allow future generations to enjoy the beauty of the stars, and children will be inspired to learn and perhaps to enter fields of science.

What is good lighting?

Good lighting does its intended job well and with minimum adverse impact to the environment. Good lighting has four distinct characteristics:

i) it provides adequate light for the intended task, but never over-lights.
Specifying sufficient light for a job is sometimes hard to do on paper. Remember that a full moon can make an area seem quite bright. Some modern lighting systems illuminate an area to 100 times as bright as does the full moon! Brighter is not always better, so try to choose lights that will meet your needs without illuminating the neighborhood. If you can't decide what to do, consulting a good lighting designer is usually your best bet.

ii) it uses fully shielded lighting fixtures, fixtures that control the light output in order to keep the light in the intended area.

Such fixtures have minimum glare from the light-producing source. "Fully shielded" means that no light is emitted above the horizontal. High-angle light output from ill-designed fixtures is mostly wasted, doing no good in lighting the ground, but still capable of causing a great deal of glare. Of course, all the light going directly up is totally wasted.

Fully shielded light fixtures are more effective and actually increase safety, since they have very little glare. Glare can dazzle and considerably reduce the effectiveness of the emitted light.

iii) it has the lighting fixtures carefully installed to maximise their effectiveness on the targeted property and minimise their adverse impact beyond the property borders.

Positioning of fixtures is very important. Even well shielded fixtures placed on tall poles at a property boundary can cast a lot of light onto neighboring properties. This "light trespass" greatly reduces and invades privacy, and is difficult to resolve after the installation is complete.

Fixtures should be positioned to give adequate uniformity of the illuminated area. A few bright fixtures (or ones that are too low to the ground) can often create bright "hot spots" that make the less lit areas in between seem dark. This can create a safety problem. When lighting signs, position the light above and in front of the sign, and keep the light restricted to the sign area; overlit signs are actually harder to read. Buildings ought to be similarly lit to offer an attractive, safe environment without overkill.

iv) it uses fixtures with high-efficiency lamps, while still considering the color and quality as essential design criteria.
High-efficiency lamps used for lighting not only save energy – which is good for a cleaner environment – but reduce operating costs. Most high-efficiency lamps last a long time, reducing costly maintenance. Highly efficient fixtures usually cost more initially, but the payback time is very short, and such fixtures will save you lots of money in a short time.

Balancing against high-efficiency, though, is the quality of the light emitted. In some applications, the yellow light cast by low-pressure (LPS) or high-pressure (HPS) sodium lamps may not be as desirable as a less efficient but much whiter compact fluorescent, metal halide, or even incandescent light source. In other applications, color is not of importance and LPS or HPS lamps do a very good job at very low cost. Well designed shielded lights can usually be lower in wattage, saving even more energy and money. They will actually light an area better than unshielded lights of higher output, because they make use of all the light rather than wasting some (or much) of it.

Why are these characteristics so important? How do they factor into a design?

- Good lighting means that we save energy and money, and we avoid hassles. A quality lighting job makes a "good neighbor". And we have a safer and more secure nighttime environment.
- Always remember that lighting should benefit people. Controlled, effective, efficient lighting at a home or business will enhance beauty, while providing visibility, safety and security.

- Poorly installed, bright lighting is offensive and gives a very poor image.

Some thoughts on cost: money talks!

- There are many cheap lighting fixtures available from most discount warehouses stores and from electrical suppliers. Are these good deals?
- Most cheap fixtures have poor control of the light output and produce a lot of glare. It usually takes better and more costly internal reflectors to get light out without glare and to give better light distribution. Modifying installed fixtures to reduce glare, or installing more fixtures to give better coverage, can be expensive.
- Cheap fixtures often have inefficient lamps and short lamp life, so they use far more energy than needed. Paying for more electricity than needed is expensive, as is the higher maintenance cost of these so-called "cheap" fixtures.

Some basic considerations

- Always remember that lighting should *benefit people.* Controlled, effective, efficient at your home or business will enhance the surroundings and give a sense of safety and security. People don't appreciate poorly installed, overly bright lighting.
- Check your site at night before installing lighting and note the existing light levels. If the area has low levels of lighting, then modest levels of light will work well for you and will fit more hospitably in the neighborhood.
- Try to keep the lighting uniform, and *reduce glare* as much as possible. Lights that make bright "hot spots" and ones that have glare make it hard to see well, especially for older people.
- Be aware that light fixtures can have different lighting patterns. Some patterns are long and narrow light cones, while others are more symmetrical. Some fixtures have internal adjustments that can change the lighting pattern to a modest extent. Pick the right pattern for your job.
- Consider using lights that turn on by motion detection. Not only will you reap big savings in operating costs, but you will have a far more effective security light due to its "instant-on" characteristics. Note that these lights can also be turned on manually. These fixtures are not expensive, and use very little energy. Higher-priced motion detection units will prove more reliable.

Appendix 4

Extracts from Articles on the Legal Aspect of Light Pollution (reproduced with permission)

Extract 1

In her article "Light Pollution: a Review of the Law" (*Journal of Planning and Environment Law*, January 1998), lawyer Penny Jewkes wrote:

> ... There are several parallels to be drawn between light pollution and noise, which occupied a similarly uncertain territory prior to 1960, when the Noise Abatement Act reflected various byelaws used by local authorities to deal with local noise problems. However, it was not until the Control of Pollution Act 1974 that effective and comprehensive controls were introduced. Light has the potential to cause distress and is an equally insidious pollutant. Noise and light are both intangible and ephemeral, yet susceptible to measurement; their detrimental effects are relatively easily avoided or stopped and both are closely associated with the development pressures of post-industrial societies. It is the perception of the relative degree, frequency and effect of the problem which causes noise pollution to be more regulated than light pollution, rather than any technological differences.
>
> ... There is significant capacity within the planning system to influence the design and installation of lighting schemes, but it has a limited ability to control the problems caused by poor lighting which is unrelated to new development. Development plans and supple-

mentary planning guidance may regulate the lighting considerations arising out of any new proposals, but the development control process is constrained by the fundamental problem that many of the lighting installations which cause this form of pollution fall outside its statutory scope. Planning permission is usually required for the carrying out of any "development" of land. As is well known, this can take two forms, namely "the carrying out of buildings, enginering, mining or other operations, in, on, over or under land." It is a question of fact in each case whether a lighting installation amounts to a building or an engineering operation. These terms are given their ordinary meaning although the courts have said that engineering operations are usually those undertaken by engineers, which would include specialist lighting engineers. Large scale installations, such as the lighting of a football stadium or public tennis courts, are clearly a form of development which comes within the statutory definition. More difficult, however, is the smaller scale lighting installation, which is probably outside planning control unless it materially affects the external appearance of a building. This qualification has to be interpreted in the light of the individual case.

... The impact of lighting on amenity and on the environment are material considerations in the decision-making process ... As a general rule, local planning authorities are not encouraged to duplicate controls imposed by other statutory bodies (such as the Environment Agency). But environmental considerations are material, and, since light pollution is not specifically controlled under any other legislation, the problem of duplication of control does not arise. The protection of a group of individual interests, such as disturbance to neighbours, is an aspect of the public interest and capable of being a planning consideration. Light is not specifically included in the list of potential nuisances. Nevertheless, some local authorities have served abatement notices under Section 80 in respect of light nuisances. It is debatable whether such a course would survive a challenge on appeal, although it might be possible to argue, in appropriate circumstances, that intense building luminance amounted to "premises in such a state as to be a nuisance".

... The common law action for nuisance is the most common method of asserting an environmental claim. Nuisance takes two forms: public and private. a private nuisance arises from a substantial interference with an individual's use and enjoyment of his property, and an action can only be pursued by the individual whose rights have been affected ... A public nuisance is one "which materially affects the reasonable comfort and convenience of a class of Her Majesty's subjects". This tort shares many of the characteristics of private nuisance. However, a public nuisance is a criminal offence and the action may be brought by the Attorney General (or the Local Authority). An individual may also bring proceedings if he has suffered some special damage over and above that suffered by the general public. Where floodlights from a sports stadium affect a large number of people living in an area, and have a particularly detrimental effect on adjacent landowners, the complainants may have an action in private and public nuisance. This type of action, in which excessive lighting is alleged, requires the courts to effect a delicate balancing exercise between neighbours' competing uses of land. The court will take numerous factors into account, such as the locality where the complaint has arisen, how often the activity occurs, whether one neighbour is more sensitive than normal and whether the other party has a good reason for carrying out the activity complained of.

Conclusion

Light pollution is a relatively newly identified environmental problem and in some of its manifestations it may appear to be a relatively insignificant issue. But insignificant to whom?

A person who is unable to sleep because of his neighbour's security light may not regard this as a minor problem. He may be perplexed by the logic which says that if his neighbour's dog keeps him awake at night the local authority environmental health officer has powers to intervene on his behalf, but he cannot do so if the insomnia is caused by glaring lights. A sensitively designed development may be absorbed into a rural scene without detriment to visual amenity and be acceptable to local people. The position might be quite different when they see it lit up at night. We want our children to inherit a planet rich with diverse species and plant-life; is it any less important that they should be able to see the Milky Way or the tail of the Hale-Bopp comet? Does it make any difference whether the survival of a rare insect is jeopardised by a light source or by a pesticide? These are some of the many questions which the problem of light pollution raises and which are only in part addressed by the existing regulations.

Extract 2

In his article "And God Divided the Light from the Darkness – Has Humanity Mixed Them Up Again?" (*Environmental Law and Management*, January 1997), law lecturer Martin Morgan Taylor wrote:

> The increase in national and local authority maintained night lighting is arguably treating the symptoms of the societal disease which manifests itself as crime: for example, increasing lighting because there is an increase in crime. The money spent on reducing the fear of crime, and on giving the impression that money is being spent on run-down areas, could be put to much better use by fighting crime at its roots ... The second matter usually cited to defend the general increase in lighting is safety. It must be accepted that without lighting, the world would not be safe at night. However, a balance is what is needed, not an absolute flood of lighting for lighting's sake ... the American "Green Lights" programme, launched in 1993, is a government backed movement to replace light fittings both public and commercial with fittings which are more economical and less ecologically harmful. The aim is to cut energy bills for participants, and reduce global environmental impact. The Environmental Protection Agency estimates that if "green lights" were implemented on all participants' land, $16 billion would be saved. This equates to 12 per cent of U.S. utility carbon/ sulphur/nitrogen dioxide emissions.

Possible solutions

It is proposed that there are solutions to the lighting problem. Firstly, there should be some form of legislation which restricts power consumption and output of domestic exterior floodlighting. There is no positive gain to be had by installing a 500 Watt light in a small back yard, but there are disadvantages, both to the environment and to neighbours. All light fittings should be properly designed, perhaps to a BS standard, so that they may point straight down and still permit the functioning of a trigger mechanism. Instructions for their installation should counsel the installer to angle the light sensibly, so that a burglar may be seen by a passer-by who will not be dazzled, and so as not to interfere with neighbours. Environmental health departments should be granted the authority to order the repositioning of poorly placed lights that interfere with neighbours or pose a risk to road traffic. Should a person not comply, then local authorities should have the power to remove the light fitting, or where the offender has been malicious or uncompliant, to prosecute ... maintaining safe levels of night-time lighting, with the intention of reducing environmental harm ... such a policy will reduce the emission of greenhouse gases and also help protect the environment. It is argued that this time has now come.

Appendix 5
Some Lighting Myths (reproduced by kind permission of Dr David Crawford, IDA)

1. The more light the better: "The more light the better" is the same type of reasoning as saying the more salt on your food the better, or the more fertiliser the better, or the more medicine the better. Obviously, there comes a point where you can have too much of a good thing. Eventually, it becomes wasteful or even harmful. Nighttime lighting is that way. We need well-lit streets, security lighting, and parking lot lighting. However, we do not need glare, clutter, confusion, light trespass, light pollution, and energy waste. Excessively bright, numerous, unshielded lights cause all of these things.

The amount of light you need depends upon the task. For example, you use low wattage colored bulbs for Christmas tree lights, and perhaps a 60 watt bulb for a porch light. If more light is better, why are night lights in a bedroom dim instead of bright? The next time you are at an airport at night look at the brightness of the taxi lights (blue color) or the runway lights (white color). They are relatively dim so as to not harm the pilot's night vision and cause confusion. Even the rotating airport beacon is not especially bright. The strobe lights on tall chimneys and radio towers are of low wattage, yet they are visible for miles. Those who claim "the more light the better" often are salespeople or manufacturers who pander to people's misconceptions to make a quick sale rather than educate their customers about truly effective and environmentally responsible lighting.

2. Light pollution only affects astronomers: Light pollution affects all of us. It robs the professional astronomer of his or her livelihood and hinders the amateur's enjoyment of their hobby. It deprives us all of one of nature's grandest wonders – the night sky. Many persons who claim this is of no importance have never gone far out of town to see what they are missing. Those who grow up in an urban environment may never see the Milky Way. How can someone miss something he has never seen? The loss of this part of nature desensitises us to other insults upon the environment. This is like saying the loss of a virgin forest is of no concern because most people won't get to see it anyway, and there are plenty of trees for lumber. The loss of wildflowers, polar bears, wolves, whales, and other threatened species, to be honest, won't affect the average person. Their loss only directly impacts biologists, or those more in tune with the natural environment than in the environment we humans create. After all, humans have done very well without mammoths, mastodons, and passenger pigeons. However, no one supports the extinction of magnificent animals. Why should we permit the loss of our skies? Not only does light pollution dim the stars for the astronomer, but it dims them for all persons. Everyone has a right to the stars.

Light pollution takes away one of our most ancient heritages and it represents visible destruction of the environment in several ways: the dome of light hanging over most cities blots out the stars; electricity is generated and wasted to light the night sky – light needs to be on the ground not up in the sky; the wasted electricity represents wasteful burning of coal, oil, and natural gas; the by-products of these wasteful burnings show up as acid rain, smoke, and carbon dioxide emission; strip mining and underground mines ravish the land to produce the coal for the wasteful burnings; runoffs from this mining pollutes rivers and streams. Thus, light pollution does far more than inconvenience a few astronomers. It is a most harmful assault upon our environment. It affects us all, and all of us ought to be concerned about it.

3. Just go out of town away from the lights: This is equivalent to saying why worry about the loss of trees and flowers in our cities. Why have urban parks? Just go out of town to see some grass, flowers, or trees. It shouldn't be necessary to go out of town to see these. If we can't have enough sense to plant trees, shrubs, and flowers all around our cities, we can at least have enough sense to plan for parks and preserve those green areas left. Why not have the same attitude toward dark skies? We are not asking people to turn off their lights. We are asking them to shield the lights, use proper wattage for the task, and turn off unneeded lights. In any event, it is no simple task to get away from the lights. Urban sky glow, the dome of light hanging over all cities of any substantial size, extends for miles and miles. For example, it easy to see the sky glow of Phoenix, Arizona, from more than 100 miles away. The sky glow from Los Angeles, California, is visible from an airplane 200 miles away. How many dark spots are left in the urban corridor in the North-eastern part of the United States? Even in the most remote portions of North America, there are dusk-to-dawn lights blaring into the darkness. The light from even one of these causes significant light trespass a mile or more away. I challenge anyone reading this to find a mountain top or plateau in the continental United States where there is no trace of light pollution visible somewhere on the horizon.

4. It's too late to do anything about light pollution: there are too many lights: This is a frequent response when I ask people why they are not more active in the light pollution struggle. It's a tough response to adequately address. Yes, the problem is enormous, growing in many areas, and very difficult to grasp fully. This doesn't mean it isn't worthy of effort. We have barely begun to fight. Just because we have a very big problem on our hands and presently few resources to bring to bear, doesn't mean we can't ultimately win. It's way too early in the struggle to say it's impossible to do anything about light pollution. Only recently has a small fraction of the public and astronomical community awakened to the problem. Only recently have we realised there are solutions to most lighting difficulties. There are now excellent fixtures available for all lighting needs. This is one of those few problems whose solution is eminently sensible, available, and which saves money in both the short term and the long haul. If you expect to rid a city of its sky glow in the next year, then you will be very disappointed. If you want to get rid of local sources of light trespass, such as a dusk-to dawn light next door or an unshielded street light on the corner, then you have a very good chance of accomplishing your goals with persistent but not obnoxious effort. You also have a reasonable chance for changing laws and instituting proper lighting techniques in your community. Over a long period good lights will replace the bad and the ugly ones. There will be a gradual slowing of the loss of dark skies and then an actual darkening of the sky in some areas. This will not happen quickly but it is possible. It will take incredible amounts of work and determination but it can be done.

4. Low Pressure Sodium (LPS) causes headaches: This is just one of hundreds of ill-founded rumors about LPS lighting. Low pressure sodium is the most energy efficient lighting available. LPS is favored by professional astronomers because it is an essentially monochromatic light source, more easily filtered out than other light sources. It produces a bright, yellow light to which the eye is very sensitive. Therefore, it is very good for street lights, parking lot lighting, and security lighting. Ask those in San Diego, San Jose, Long Beach, and Glendale, Arizona, where LPS is used extensively. Why isn't it used more often? The answer is complex. Several large lighting manufacturers do not make LPS fixtures or bulbs and campaign against it. It has no color rendition, which bothers many persons, especially when they first see it, and it should not be used for any lighting application that needs good color. LPS fixtures and ballasts are expensive and not readily available, even though LPS use quickly saves money. LPS lighting does not produce headaches any more than any other type of outdoor lighting. In fact, it tends to produce less glare than mercury vapor lights or high pressure sodium (HPS) lights and is thus probably less likely to give headaches. LPS bulbs are no more dangerous to dispose of than any other type of light bulbs. In fact, consider the toxic substances that are found in other bulbs. Mercury vapor lights contain mercury. In the metallic form, mercury is not especially toxic but many of its salts are quite poisonous. HPS bulbs contain metallic sodium just like LPS bulbs; therefore, they have the same disposal problems as the LPS bulbs, mainly the metallic sodium which is highly reactive. If HPS or LPS bulbs are carefully broken under water, the sodium reacts with the water to give sodium hydroxide, everyday lye, the same substance as in drain cleaners. How about all the glass? Well, this is a problem with disposing of any light bulb. Metal halide bulbs contain all sorts of toxic metallic salts. The bottom line is that the disposal of a large number of light bulbs is an environmental problem no matter what the bulb type.

6. Security lights prevent crime: Does outdoor nighttime lighting prevent crime? The answer is: nobody knows. In some cases, lighting seems to deter crime and it makes people feel more secure, but in reality they may be just as secure without the lighting. In some cases, lighting probably increases crime because it draws attention to a house or business that would otherwise escape attention. Most crimes, violent and otherwise, take place during the day. After all, criminals need light to do their work, too. A dusk-to-dawn light shining all night in a rural area probably is an inducement for robbery and vandalism. A passer-by might not otherwise notice that the farmhouse is even there. An infrared motion-sensor security light which comes on only when someone steps into the beam makes a lot of sense. It is only on when needed, thereby conserving energy. Its sudden illumination serves to frighten away the criminal. These lights are now beginning to replace some of the all-night dusk-to-dawn 175 watt mercury vapor lights. This makes good sense from the economical, environmental, and crime prevention points of view. The motion-sensor security lights can cause light pollution and light trespass if too high a wattage spotlight is used, or if they are not aimed down toward the ground. They should also have some shielding. Do street lights, parking lot lights, and security lights prevent crime? Maybe yes, maybe no. If they are overly bright with much glare, they actually make it easier for a criminal to hide in the deep shadows produced by objects in the harsh glary light and encourage crime rather than discourage it. Well-lit streets with even, uniform lighting, low glare, and utilising fully-shielded fixtures probably have lower vehicle and pedestrian accident rates. How about bright lights in a parking lot? How many people do you know whose car has been broken into during the day, or while directly underneath a light at night? One speaker at a recent lighting symposium recounted how his car was robbed at a local mall. It sat near a store entrance and was directly under a bright light! There are simply no good scientific studies that convincingly show the relationship between lighting and crime. Our cities are far more brightly lit than ever. Yet, the crime rate soars. Maybe lights directly lead to crime. Who knows? One study at a small eastern college showed almost all violent nighttime crimes took place in well-lit places. This study, while informative, cannot be generalised to other locales because of the somewhat unique nature of the college and the college town. Crime is a very complex sociological phenomenon controlled by many factors, and it will vary considerably from place to place. My own personal opinion is that crime is little affected by nighttime lighting for better or worse. Main arterial streets should be well lit to reduce automobile and pedestrian accidents. Busy malls should have good lighting to reduce accidents and perhaps deter crime. After business hours this lighting can be reduced or even turned off. Security lighting can be at a relatively low level. This saves money, and not much light is needed to find your way to a door or find your way out to your car. Not much light is needed to see a suspicious-looking person loitering around. No matter what the lighting situation, the proper wattage, not overkill, should be used, and all light should come from full-cut-off, shielded fixtures. Low pressure sodium lighting is ideal for many of these applications because of its very low operating cost.

7. Only astronomers care about light pollution (those persons fighting light pollution are just crazy idiots): Anyone who takes a well- educated and reasoned approach toward environmental or quality of life issues is not a "crazy idiot". We (and many others as well) are concerned about light pollution, light trespass, radio pollution, and space debris. After all, the night sky is part of everyone's environment, enormous amounts of energy are wasted lighting the night sky, radio astronomers have to struggle to find usable portions of the electromagnetic spectrum for their work, and space debris is a rapidly growing problem for spacecraft (and people) in orbit. Why should someone be considered a nut because he or she is concerned about the environment? However, persons involved in environmental causes must carefully define the problem they want to solve, learn the facts, appreciate the legitimate perspective of their opponents, and offer people solutions rather than complaints. This is IDA's philosophy and modus operandi. Light and radio pollution are solvable problems if the facts are properly conveyed to the public. Light pollution is the one form of pollution whose solution immediately saves money. Not just astronomers care about light pollution and light trespass. IDA's Board of Directors consists of a physician, a lawyer, two lighting designers, a city streetlighting director, as well as professional and amateur astronomers. Many IDA members are not astronomers or even particularly interested in astronomy. They are concerned about energy conservation, preservation of our environment, and proper nighttime outdoor lighting. They include homemakers, scientists, lawyers, pilots, doctors, engineers, retired persons, and so forth. Much of IDA's strongest support comes from professional lighting engineers, lighting suppliers, and lighting manufacturers.

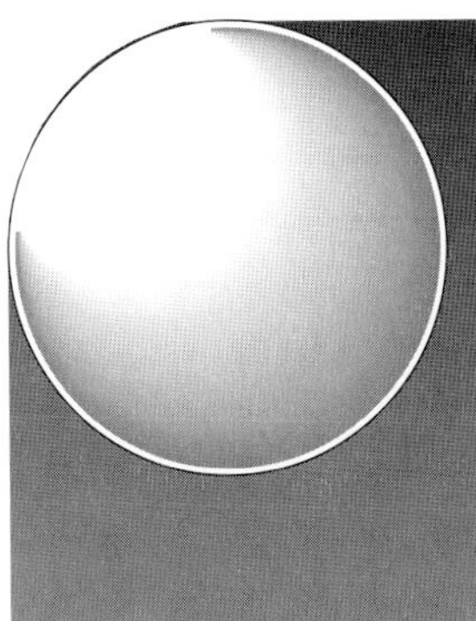

Appendix 6
Advice from IDA and CfDS on Seeking Action on Lighting and the Environment from Local Authorities and Others

IDA's standard letter informs and suggests courses of action:

> Dear … ,
> Here is an environmental issue you might not have heard of yet: Light Pollution. It is a growing threat to our night-time environment, one that has already seriously harmed astronomers, both amateurs and professionals. We are faced with the distinct possibility that in only a generation or two very few people will be able to have a "live" view of the universe. Urban sky glow will have blotted out the dark sky, just as a lighted room blots out the view of a slide show.
> Components of light pollution include:
>
> 1. Urban Sky Glow: it is destroying humanity's view of the universe.
> 2. Glare: blinding us and harming visibility; Glare is never good.
> 3. Light Trespass: someone's outdoor lights offending us, "trespassing" on our property.
> 4. Clutter: trashing the nighttime environment, and causing confusion as well.
> 5. Energy Waste: wasted light costs over one billion dollars a year, in the USA. alone.
>
> There are solutions to all of these problems. Quality lighting is the key. These solutions preserve the dark

skies, improve the quality of the nighttime lighting and the nighttime environment, and save money as well. It is a Win/Win/Win situation.

Awareness of the problem and of the solutions is needed, of course, but is often lacking, even among lighting professionals. Lack of awareness (rather than resistance) is the main problem in implementing these solutions.

You can help. Please do. Here's how.

First, become aware. Insist on quality lighting. Use it yourself. Quality lighting is well shielded (so the light is used, not wasted), uses the right amount of light (not overkill), includes time controls when possible, and includes the use of low pressure sodium (LPS) as the light source when possible. (LPS is the most cost-effective light source, excellent where color rendering is not critical.) Quality light is directed downward where it is needed, not up or sideways where it is wasted and causes glare, light trespass, and bright skies.

Second, a non-profit organisation, the International Dark-Sky Association (IDA), is very active in raising awareness of the issues and in pushing for solutions to the problem. The IDA also addresses the related issues of radio interference, space debris, and other environmental threats to our view of the universe. All of these problems adversely affect the general public and seriously threaten the future of frontier astronomical research everywhere on Earth.

(Included are several sheets which discuss this issue in greater detail.) I hope you will take the time to read them, and to think about the issue. The IDA, and all who care about the environment and our quality of life, need your help. Please become aware of wasteful lighting, and do what you can to help.

Sincerely yours,
Etc.

CfDS sends the following advice to people seeking guidance on countering problems caused by stray light.

If writing to your council, and in spreading awareness about sane lighting, you might consider the following options:

(1) Find out who is responsible for lighting. If "A" class roads are lit, it is normally the Highways Agency; minor roads and side streets are normally lit by the county or district council. All councils will have a lighting engineer, and (s)he should be following the guidelines of the Institution of Lighting Engineers (ILE), which recommend minimal upward light. All

major lighting companies, dozens of local councils, the Institute of Environmental Health Officers, the Council for the Protection of Rural England (CPRE – joint producers with the BAA of the *Starry Starry Night* leaflet), and many other bodies agree with the BAA/CfDS and the ILE that light pollution is a problem to be confronted. Anyone installing glary roadlights with upward waste light is simply behind the times and not environmentally aware.

(2) Refer your lighting engineer(s) to the ILE, and also to the many lighting firms now producing full-cut-off and semi-cut-off lamps. With the latter, there should still be minimal upward light (what lighting people call Upward Waste Light Ratio (UWLR)).

(3) Tell friends and neighbours about light pollution, using CfDS material. An astronomer in a nearby back garden at night is a far more effective security device than any number of 500 W lights! Try to let everybody (local press?) know that the environment above is just as valuable to the human spirit as that below. Get your local environmentalists involved: the CPRE, for example, has produced a "light pollution charter", and declares itself firmly committed to eradicating local light pollution;

(4) Resist arguments such as: "*it's not the lights shining upwards. It's mostly reflection off the ground*" – anyone standing on a hill over a large conurbation can see with their own eyes that it is the lamps which are glowing brightly, not the ground! Or "*cut-off lights are more expensive*" – they may be, but what price the environment? The trend is towards environmentally friendly lighting, and councils with glary lights may well have to replace them with something better early in the next century. Why not do it now and save money later? Or "*you need more lights with FCOs as they have to be spaced more closely together*" – no they don't. The local street lights in the CfDS coordinator's area have now been replaced with FCO and SCO designs. The night sky is much improved (which further refutes the ground reflection argument) and they are on exactly the same columns as the old, wasteful lights. The M5 motorway was recently relit with FCOs – four FCOs for every five old, glary lights, so they are actually now further apart. It's column height and the optical reflectors in the lamps which control the light spread, not the distance between them. Or "*lights have got to be bright to defeat crime*" – there is no proof that light and crime are related. Some studies show a reduction in crime where lighting has been introduced or upgraded. Other studies show the opposite, or no change. The vast amount of crime which takes place in broad daylight suggests that ambient light levels do not deter criminals. The best friend of the modern burglar is the sideways-pointing 500 W "security"

light, which emits a wall of dazzling glare behind which he can work unseen.

CfDS wishes you success in spreading the message that effective lighting and an attractive natural night-time environment are not mutually exclusive.

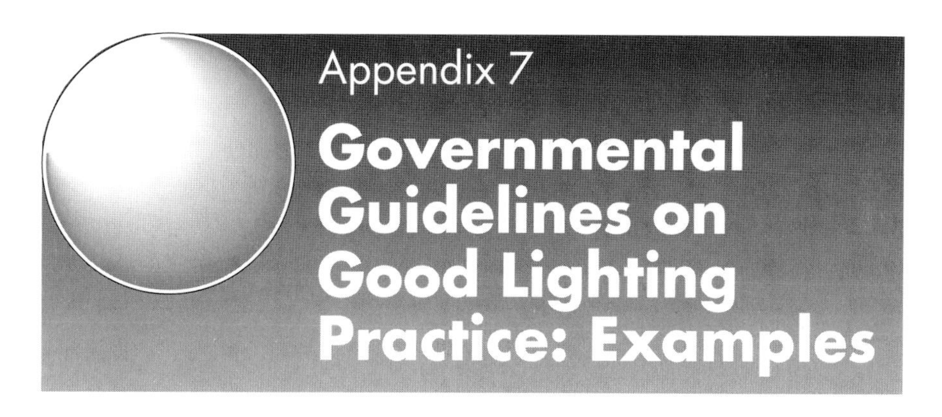

Appendix 7
Governmental Guidelines on Good Lighting Practice: Examples

Guidelines issued by national governments for the reduction of light pollution are few; for example, I have been unable to discover any American governmental agency guidelines on good lighting (if any appear, the IDA website will undoubtedly feature them). The first such advice was probably that contained in the document *Lighting in the Countryside – Towards Good Practice* (ISBN 0 11 753391), issued by the British Government's Department of the Environment in 1997, in collaboration with the UK Countryside Commission and with input from CfDS, CPRE, ILE and the Royal Fine Arts Commission. Though restricting its brief to rural lighting, this 80-page guide contains advice on all its aspects. In its section on "Action on Lighting in the Countryside" we read:

> Lighting in itself is not a problem; it only becomes a problem where it is excessive, poorly designed or badly installed. Better use of the planning system to influence lighting proposals; better awareness of the potential adverse impacts of light amongst developers, manufacturers, retailers and the general public; and improved lighting design and landscape design are among the most important ways of tackling issues of overlighting ... For all but the simplest lighting scheme, professional advice whether from the lighting manufacturer or from a qualified lighting engineer/designer, is recommended. The range of lighting standards and lighting products on the market today is very broad ... If action on lighting in the countryside is to be effective it will require the close co-operation and participation of all those involved in planning, designing, and installing lighting schemes. The responsibility for tackling lighting issues is very much a shared

> one ... Local authority planners should recognise the cumulative impacts of lighting on countryside character, and be more pro-active. They should consider the need for policies on lighting in the development plan ... Developers should look differently upon lighting than they did in the past, and should not automatically assume that it is a good thing. This implies a more critical assessment of lighting need and alternatives, and a greater willingness to consider the removal or upgrading of intrusive lighting. In judging the costs of lighting they should take a long term view and give due weight to energy and maintenance costs as well as capital costs ... Lighting engineers and designers should adopt a more structured approach to assessing the environmental impacts of lighting installations... Manufacturers and suppliers of lighting equipment should provide a design service that is as impartial and responsible as possible, and should focus increasingly on high quality lighting products ... In relation to security lighting that is intended for DIY installation, retailers have a special responsibility to ensure that good information is available on how to choose appropriate equipment, minimise light levels, and control light pollution through good installation ... Lastly, members of the public have a vital rôle in the control of light pollution. They are responsible not only for most domestic security lighting, but also for much of the small scale lighting on commercial and business premises that does not need planning permission. They should take great care in the selection and installation of lighting equipment, and if in doubt, should always seek professional advice. (Crown copyright is reproduced with the permission of the Controller of Her Majesty's Stationery Office.)

In 1998, the Environment Agency of Japan, in collaboration with the CIE Japanese section, published its *Guidelines for Light Pollution: Aiming for Good Lighting Environments*. In his introduction to this minutely detailed 93-page document, Syuzo Isobe, of the National Astronomical Observatory at Mitaka, Tokyo, describes the necessity for the intervention of lighting engineers on a larger scale, good monitoring and the increased use of satellite technology in the measurement of wasted energy. The document recommends close inspection of proposed lighting by the use of standardised checking procedures for all lighting tasks, and lists procedures to be undertaken. In the body of the text, we read:

> Checklist for outdoor lighting:
> Targets: facility maintenance companies, facility managers, designers of environmental lighting, and citizens.

When installing lighting that takes into consideration the lighting environment ... Studies for sufficient and efficient lighting are necessary. For these studies, it is important to clearly identify the objectives for each individual lighting installation, and this is related to the suppression of spill light or obtrusive light, and attaining a more efficient lighting facility. Check procedure:

Preparation of an "overall lighting plan":
(1) understanding the type of facility (e.g. residential, business, public);
(2) selection of lighting group (type of function, e.g. transit, crime prevention, decorative);
(3) understanding the surrounding environment;
(4) arrangement of compatibility of lighting groups;
(5) preparation of "overall check sheet" and "lighting group arrangement plan".

(Reproduced by kind permission of Syuzo Isobe.)

Glossary of Terms

Words and phrases it might be useful to know when discussing light pollution:

ambient light The total light level or effect, or amount of light perceived, in one's surroundings.

baffle A plate inserted within or just outside a luminaire to shield the light from direct view.

ballast Electrical devices used in conjunction with a discharge lamp to start and control it.

candela (cd) or **standard candle** The SI unit of luminous intensity.

CfDS The British Astronomical Association's Campaign for Dark Skies. See Appendix 1 and Bibliography.

CIE The Commission Internationale de l'Eclairage (International Lighting Commission), based in Vienna. See Bibliography.

colour rendering/rendition The perceived effect on objects of different colours of lights of different types.

colour temperature This term describes the actual colour of the light source itself, as opposed to that of the light issuing from it.

column The post upon which a lamp is mounted.

cones See **rods and cones.**

dark adaptation The transition of visual processes within the eye to darker surroundings. **See rods and cones.**

disability glare (veiling luminance) Glare causing reduced visual performance. See Section 1.2.

discomfort glare Glare producing discomfort or annoyance without necessarily interfering with visual performance. See Section 1.2.

FCO This stands for full-cut-off, referring to a lamp with a flat glass panel beneath, which, when mounted horizontally, emits no light above the horizontal. See also **SCO.**

fluorescent A long-life, relatively cheap whitish light source based on a gas discharge process, where electrons pass through a tube and interact with a phosphor coating.

flux Luminous flux is the rate of flow of particles of light energy, measured in watts or ergs/sec.

fovea A small central depression in the back of the retina containing cone cells: the area of sharpest vision.

full-cut-off See **FCO.**

IDA The International Dark-Sky Association. See Appendix 1 and Bibliography.

IESNA The Illuminating Engineering Society of North America. The USA's professional guidance body for lighting engineers. See Bibliography.

ILE The Institution of Lighting Engineers. The UK's professional guidance body for lighting engineers. See Bibliography.

incandescent Describes a light source based on electricity passing through a thin filament (usually tungsten) which glows brightly.

kWh Kilowatt hour: unit of energy equal to the work done by one thousand watts of power acting for one hour.

light spill The emission of light outside the premises which the lighting is supposed to illuminate.

light trespass Troublesome light entering areas or premises outside the boundary of the premises to be illuminated.

lumen (lm) The SI unit of luminous flux, being the flux emitted in a solid angle of 1 steradian by a point source with uniform intensity of 1 candela (q.v.).

luminaire A word not found in many dictionaries, but widely used in the lighting community to denote the lamp and its surrounding casing and optics.

lux (lx) The SI unit of illumination, being a luminous flux of 1 lumen (q.v.) per square metre. The value for the full Moon is about 0.2–0.3 lux.

photometry The measurement of the level and distribution of light.

reflectance The amount of light reflected by a given surface (the ratio of the reflected flux to the incident flux).

reflectivity The ability of a surface to reflect radiation (technically, equal to the reflectance of a layer of material sufficiently thick for the reflectance not to depend on the thickness).

rods and cones Cells in the retina of the eye. Rods are cylindrical cells containing rhodopsin ("visual purple"), and are sensitive to dim light but not to colour. Cones are conical cells which are sensitive to colour and bright light. The process of dark adaptation involves the rods taking over visual duties from the cones. Interestingly, there are no rods in the centre of the fovea (q.v.), which explains the astronomer's "averted vision" trick (objects appearing more distinct if you look slightly to one side of them).

SCO Semi-cut-off: a lamp type which has a shallow bowl beneath, and emits little or no light skywards.

SON Another name for high-pressure sodium sources (see Section 1.3).

SOX Another name for low-pressure sodium sources (see Section 1.3).

skybeam A concentrated beam of light sent into the sky deliberately, usually for the purposes of advertising (often erroneously called a "laser").

skyglow Unwanted light emitted into the night sky from poorly aimed lamps.

stray light See **light spill.**

street furniture All manufactured items commonly seen along roadsides, e.g. lighting columns, telephone poles.

veiling luminance See **disability glare**.

visibility Clarity of vision; how well we see something. The purpose of a good light should be to increase visibility: to reveal and not conceal.

UWLR, ULR, upward flux The abbreviations stand for upward (waste) light ratio. All these terms refer to the relative amount of the light emitted above the horizontal.

Bibliography

Organisations

British Astronomical Association (BAA)/**Council for the Protection of Rural England** (CPRE):
Starry Starry Night (April 2000)

BAA Campaign for Dark Skies (CfDS):
Domestic and Commercial Security Lighting
CfDS Newsletter (twice yearly)

The CfDS website has a light pollution reading list:
www.dark-skies.freeserve.co.uk
All BAA/CfDS publications available from BAA, Burlington House, Piccadilly, London W1J 0DU, UK. (0)207 734 4145.
E-mail: office@baahq.demon.co.uk

British Standards Institution:
British Standard (BS)5489 (Code of Practice for Road Lighting)
Available from the BSI, 389 Chiswick High Rd, London W4 4AL, UK. (0)208 996 9001.

Chartered Institution of Building Services Engineers/ Institute of Light and Lighting:
Lighting Guides (LG1 Industrial, LG4 Sports, LG6 The Exterior Environment)
Available from CIBSE/ILL, 222 Balham High Rd, London SW12 9BS, UK. (0)208 675 5211.

Commission Internationale de l'Eclairage (CIE):
Guidelines for Minimizing Urban Sky Glow Near Astronomical Observatories
Guide for Floodlighting
Recommendations for the Lighting of Roads for Motor and Pedestrian Traffic
Guidelines for Minimizing Sky Glow
Guide to the Lighting of Urban Areas
Available from CIE, Central Bureau, Kegelgasse 27, Vienna, Austria. (001)431 714 3187.

Countryside Commission/ Department of the Environment (UK):
Lighting in the Countryside – Towards Good Practice (ISBN 0-11-753391-2)
Available from The Stationery Office, PO Box 276, London SW8 5DT, UK. (0)207 873 9090.

Department of Transport (UK):
Road Lighting and the Environment
Available from DoT Sales Unit, Government Building, Block 3, Spur 2, Lime Grove, Eastcote HA4 8SE, UK. (0)208 429 5170.

International Dark-Sky Association:
IDA newsletter (quarterly)
Available from ida@darksky.org
The IDA website has many articles, extracts and references: www.darksky.org

Illuminating Engineering Society of North America:
Recommended Practice on Roadway Lighting (IESNA RP-8-00)
Available from IESNA, 120 Wall Street, New York, NY 10005, USA. 212-248-5000.
www.iesna.org

Institution of Lighting Engineers:
Guidance Notes for the Reduction of Light Pollution
Lighting the Environment – A Guide to Good Urban Lighting
Domestic Security Lighting, Friend or Foe?
Available from the ILE, Lennox House, 9 Lawford Rd, Rugby CV21 2DZ, UK. (0)1788 576492. www.ile.co.uk

Royal Fine Art Commission:
Lighten Our Darkness
Available from RFAC, 7 St James's Square, London SW1Y 4JU, UK. (0)207 839 6537.

Books and Articles

Allen, R, *Star Names – Their Lore and Meaning* (1899), Dover, 1963 (ISBN 0-486-21079-0)
Avery, D, The Proper Use of Light Therapy, *Directions in Psychiatry*, vol. 19, 1999
Bower, J, The Dark Side of Light, *Audubon Magazine*, March–April 2000
Clark, BAJ, *Outdoor Lighting Principles for Australia in the 21st Century* on www.gsat.edu.au/astrovic
Coren, S, *Sleep Thieves: an Eye-opening Exploration into the Science and Mysteries of Sleep*, Free Press, 1996 (ISBN 0-684-82304-7) (in IDA www.darksky.org information sheet 108)

Crow, D, The Story of Light, *The Lighting Journal*, January–February 1999
Griggs, B, *Reinventing Eden*, Quadrille, 2001 (ISBN 1-902757-87-4)
Harris, J, Lighting the Queen's Highway, *The Lighting Journal*, December 1993
Henbest, N, Save Our Skies, *New Scientist*, February 1989
Isobe, S/Environment Agency of Japan, *Guidelines for Light Pollution: Aiming for Good Lighting Environments* (1998)
Jewkes, P, Light Pollution: a Review of the Law, *Journal of Planning and Environment Law*, January 1998
Light Pollution and the Law: What Can You Do?, *Journal of the British Astronomical Association*, October 1998
Kitchin, C, *Solar Observing Techniques*, Springer, 2001 (ISBN 1-85233-035-X)
McNally, D (editor), *The Vanishing Universe: Adverse Environmental Impacts on Astronomy* Cambridge University Press, 1994 (ISBN 0-521-45020-9)
Mizon, R, Heaven's Light, *Astronomy Now*, November 1999
Light Pollution, Mizar Astronomy/CfDS, 2000
Mobberley, M, *Astronomical Equipment for Amateurs* Springer, 1998 (ISBN 1-85233-019-8)
Morgan Taylor, M, "And God Divided the Light From the Darkness: Has Humanity Mixed Them Up Again?", *Environmental Law and Management*, January–February 1997
Mostert, H, A Clear-Sky Detector Using Reflected Artificial Light, *Journal of the British Astronomical Association*, vol. 93, no. 5, August 1983
Nash, D, *Filters* on www.xmission.com/~dnash
Painter, K, The Social History of Street Lighting, *The Lighting Journal*, November–December 1999
Pearce, F, Declaring a Curfew on Aurora Metropolis, *New Scientist*, March 1995
Pollard, N, Skyglow Conscious Lighting Design, *International Journal of Lighting Research and Technology*, vol. 26, no. 3, 1994
Purves, L, Light Pollution section in *Town and Country* (eds. Barnett A and Scruton R) Cape, 1998 (ISBN 0-22-405-2500)
Ramsey, M and Newton, R, *The Influence of Street Lighting on Crime and Fear of Crime,* Home Office Crime Prevention Unit, papers 28 and 29, 1991
Ratledge, D (ed.), *The Art and Science of CCD Astronomy*, Springer
Sherman, L, Gottfredson, D, MacKenzie, D, Eck, J, Reuter, P and Bushway, S, *Preventing Crime: What Works, What Doesn't, What's Promising. A Report to the United States Congress.* Prepared for the National Institute of Justice. Department of Criminology and Criminal Justice, University of Maryland at College Park, 1997.
Simpson, M, Social Factors Behind the Development Of Outdoor Lighting, *The Lighting Journal*, June–July 1995
Tonkin, S, *Astro FAQs* Springer, 2000 (ISBN 1-85233-272-7)
website (filters, etc.): www.aegis1.demon.co.uk

Some Star Atlases and Observing Guides

Burnham, R, *Burnham's Celestial Handbook* (3 vols.), Dover, 1978 (ISBN 0-486-24063-0)
Kanipe, J, *A Skywatcher's Year*, Cambridge University Press, 1999 (ISBN 0-521-63405-9)
Karkoschka, E, *The Observer's Sky Atlas*, Springer, 1999 (ISBN 0-387-98606-5)
Norton's 2000.0 Star Atlas and Reference Handbook (ed. Ridpath, I), Longman/Wiley (ISBN 0-470-21460-0: USA; 0-582-03163-X: UK)
Tirion, W, Rappaport, B and Lovi, G, *Uranometria 2000.0* (2 vols.), Willmann-Bell (ISBN 0-943396-14-X)
A list of hundreds of articles on light pollution can be found on:
http://debora.pd.astro.it/cinzano/refer/index.htm

Software

Images of Light Pollution, Chris Baddiley/CfDS (see above)
Ratledge, D, *Software and Data for Practical Astronomers*, Springer, 1998 (ISBN 1-85233-055-4)
RealSky CD, Astronomical Society of the Pacific www.aspsky.org
SkyMap Pro, Thompson Partnership www.skymap.com
The Sky, Software Bisque www.bisque.com

Index

Index of Objects in observing section 2.3